NEURAL ARCHITECTS

THE SAINSBURY WELLCOME CENTRE FROM IDEA TO REALITY

GEORGINA FERRY

UNICORN

Published in 2017 by
Unicorn, an imprint of
Unicorn Publishing Group LLP
101 Wardour Street
London
W1F 0UG
www.unicornpress.org

ISBN 978-1-910787-48-9

10 9 8 7 6 5 4 3 2 1

Written by Georgina Ferry
Designed by Kathrin Jacobsen
Printed in Wales by Gomer Press

CONTENTS

DAVID SAINSBURY

LORD SAINSBURY OF TURVILLE, HonFRS

When my Charitable Trust decided to build a new neuroscience laboratory we set ourselves the challenge of building one that would set a new standard for laboratory design. Our aim was to provide the scientists with a pleasant working environment, and maximum flexibility in the configuration of the laboratories to meet constantly changing ways of working. The layout of the building should also encourage frequent encounters between scientists and a collaborative ethos within the laboratory. And it should do all this within a building which was a beautiful addition to the streetscape.

I think that it is fair to say that in the past, few world-class architects have been asked to design laboratories, and certainly not a lot of time has been given to understanding the way that scientists work today and might work in the future. In this case, however, the architect Ian Ritchie and his colleagues, with scientific advisers, engineers and specialists, visited laboratories across the world, and consulted widely with neuroscientists. As a result, all our objectives have been met, and I am confident that the laboratory will help set a new standard for laboratory design in the UK, and probably worldwide.

My fundamental test of any building is whether it is one that I would personally like to live or work in, and if I was a young neuroscientist today I know that I would be delighted to work in the offices and laboratories that have been created. The experimental areas will also, I believe, very effectively meet the complex and demanding standards of the neuroscientists, and, therefore, significantly increase their creativity and productivity.

The building is also a beautiful and innovative one, which enhances the local streetscape into which it comfortably fits. In the future I am sure it will be known for the quality of the architecture as well as for the quality of the science done there.

Finally, I would like to thank my wife, Susie, who has taken a personal interest in the project from the start, who sat on the panel that chose the architect, Ian Ritchie, and who helped enormously in fostering the collaboration that has characterised the development of the building, and made it such a success.

SIR WILLIAM CASTELL LVO

CHAIRMAN, THE WELLCOME TRUST, 2006 - 2015

The Wellcome Trust has a deep commitment to increasing our understanding of the extraordinary apparatus of the brain. With this knowledge, we can improve health for millions.

Applied, genomic and fundamental approaches to brain research are all crucial for this human endeavour, and to increase global knowledge. We are already making important strides in genomic brain research, and Queen Square's Wellcome Trust Centre for Neuroimaging is a guiding light in applied brain research. Now the Sainsbury Wellcome Centre, a dedicated hub focused purely on fundamental brain research, will take us further towards unravelling the brain's mysteries.

Some would argue the success of the Centre will rest solely on the people it contains. But I believe strongly that if you are working in research, having an inspiring environment designed specifically around your needs has an immeasurably positive impact. Whenever I go to Oxford I am filled with wonder and enthusiasm, not just because of the remarkable academics around me, but because of the institution's impressive architecture and history.

It is increasingly true that science cannot, and should not, exist in a bubble. Beyond its close proximity to UCL to aid scientific collaboration, everyone working in the Centre will benefit enormously from its location in London's commercial creative hub. Being immersed in creative industries such as fashion, marketing and advertising will open the researchers up to other influences, expand horizons, and energise their vital research.

Having now seen the Centre, I'm sure its extraordinary design will attract equally extraordinary research talent. And I know that those who arrive there will feel a sense of privilege, but also an obligation to the people passing by its striking glass front, to push forward their work and achieve the remarkable.

'Aristotle, 2000 years ago,
was asking how is the mind
attached to the body.
We are asking that question still.'

SIR CHARLES SHERRINGTON, 1949.

JOURNEY TO THE CENTRES OF THE BRAIN

ORIGINS OF THE SWC

To the young David Sainsbury, recently arrived in Cambridge to study history, it was an intriguing challenge. His college, King's, set all its students the task of writing a general essay, and the subject for 1960 was 'Consciousness'. Not having thought about the subject much before, he began to read. The philosophical literature on the nature of mental experience left him unsatisfied. Consciousness clearly arose in the brain – but how? For enlightenment, Sainsbury turned to a little book published a decade before: *The Physical Basis of Mind* (Blackwell 1950), edited by the Cambridge historian and BBC radio producer Peter Laslett.

The book contained transcripts of a series of talks broadcast on the BBC Third Programme during 1949. The contributors, who included the 92-year-old Nobel-prizewinning physiologist Sir Charles Sherrington, were a roll-call of the great and good in neurophysiology and philosophy. 'It introduced me to the whole idea of neurophysiology, and I was rather intrigued by that', says Sainsbury. He immediately accepted the idea that mind and brain could not be separated, and decided to switch from history to natural sciences so that he could study the problem further. It was a bold choice: his years at Eton had given him a thorough grounding in classics and humanities, but barely touched on science. 'I got permission to do two years of psychology, though I wasn't remotely equipped', he says.

That eager undergraduate is now Lord Sainsbury of Turville, Chancellor of Cambridge University, former chairman of the Sainsbury supermarket chain, Minister of Science and Technology from 1998-2006 and settlor of the Gatsby Charitable Foundation. The Sainsbury Wellcome Centre for Neural Circuits and Behaviour (SWC), recently opened at University College London, owes its inspiration directly to the spark of curiosity about the brain and mind ignited more than half a century before.

REFLECTIONS ON REFLEXES
Charles Sherrington showed that reflexes, such as blinking when something touches the eye or pulling the hand back from a hot object, involve more than a simple interaction between the sensory nerve that feels the stimulus and the motor nerve that moves the muscle. They depend on activity in a circuit: for a muscle to contract efficiently, nervous signals must also cause its opposing muscle to relax.

Two individuals profoundly influenced Sainsbury's new-found passion for science: Roger Freedman and Richard Gregory. Freedman was a fellow student: the two had met within days of arriving at King's. 'He was a scientist with amazing knowledge over fields of literature and other areas', says Sainsbury. 'Up to that point I thought, in a schoolboyish way, that scientists were rather dim people. After my first two years in history I discovered that most of the interesting people were scientists.' For his part, Freedman saw Sainsbury's decision to switch to science as heroic: 'I have never seen anyone work so hard', he says.

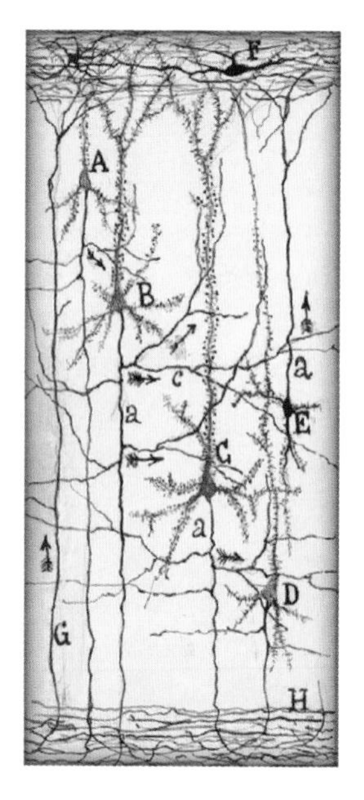

THROUGH THE MICROSCOPE
In the 19th century microscopes revealed that the brain was made up of millions of cells connected by fine fibres. Camillo Golgi and Santiago Ramòn y Cajal developed techniques to visualise single cells. Their discoveries led to the 'neuron doctrine' – the idea that nervous activity depended on signalling between individual nerve cells, or neurons.

Freedman himself pursued the physiology option within Natural Sciences for two years. 'I had the incredible good fortune to be supervised by Horace Barlow [Professor of Physiology and Fellow of King's]', he says. 'I remember reading Horace's celebrated paper about the reduction of redundancy in the visual system of the frog. It was one of those things that changes the way you see the world.' Naturally he communicated his enthusiasm to Sainsbury, whose own supervisor was a young lecturer in the Department of Experimental Psychology, Richard Gregory. Gregory was one of the most creative and enthusiastic scientists of the 20th century. 'He gave me a totally different view of what a scientist was', says Sainsbury. 'He had this wonderful laugh, and produced a never-ending stream of ideas. I got intrigued by the visual illusions he was working on.'

It was an exciting time in the field. New technology made it possible to record the activity of brain cells in anaesthetised animals and associate it with features of perception or behaviour. Only a few years previously, David Hubel and Torsten Wiesel at Harvard University had followed Barlow's lead and shown how the cortex represents the complexity of the visual world by breaking it down into simpler elements – work that won them the Nobel prize in 1981. Donald Hebb, of McGill University in Canada, had developed a theory of how neurons, organised into networks, might underlie learning and behaviour.

'I think at that time I had already understood that the interesting question was "How do neural circuits process behaviour?"', says David Sainsbury. Briefly he considered an academic career, but instead conformed to the expectation that he would join the family business. Thirty-five years later he was poised to answer the question at last: not through his own research, but by planning and funding the SWC.

John O'Keefe is Professor of Cognitive Neuroscience at University College London, and founding director of the SWC. He arrived at UCL from McGill University in Montreal, Canada, in 1970, with a new technique in his pocket. Influenced by the work of Donald Hebb at McGill, he had found a way to record the electrical activity of nerve cells in animals that were awake and exploring their environments. 'I've always been interested in how brain activity supports thinking, concepts and perceptions', he says. He focused on a particular brain region, the hippocampus, known through studies by Brenda Milner (also at McGill) on her brain-damaged patient HM to be critical for the formation of new memories. He discovered that 'place cells' in the hippocampus fired in response to the animal's location in space. With his colleague Lynn Nadel, in 1978 he published a book entitled *The Hippocampus as a Cognitive Map*, arguing that the role of the hippocampus was to provide a spatial framework for everyday experience.

'It's only recently that most people have come round to the idea', he says. They began to be persuaded when others discovered cells in other parts of the brain that communicated with hippocampal cells, while themselves firing in response to spatial triggers such as the direction in which the head was pointing, or the animal's coordinates in a particular environment ('grid cells'). 'Since then there's been a huge amount of work in computational neuroscience,' says O'Keefe, 'including people in our group trying to form models of how these all fit together to provide a mapping system that would tell the animal where it is in the environment, where it wants to go, how to get there and what to expect when it does get there.' In 2014 O'Keefe shared the Nobel Prize for Physiology or Medicine with his former colleagues, the Norwegian scientists May-Britt and Edvard Moser, for their complementary discoveries of place cells and grid cells.

'A lot of our work is recording not from single cells in conscious, awake animals but from large groups of cells', says O'Keefe. 'We believe that it's the pattern across a large group of cells that provides the code.' This work primarily makes use of two new techniques. One employs silicon wafer technology to make electrical recording probes that can record from hundreds or even thousands of cells at the same time. Gatsby and the Wellcome Trust are supporting UCL as part of consortium including the Howard Hughes Medical Institute at Janelia Farm and the Allen Brain Institute in Seattle to design, develop and test probes that will be able to record from a higher density of brain tissue than ever before. The probes are being manufactured by the silicon technology company Imec in Belgium.

The other technique, known as two-photon imaging, involves shining laser light on the hippocampus, and imaging the cells using fluorescence to label them according to their activity. 'So you can actually see the cells while the animal's doing things, and see when

MADE FOR MAZES
The 20th century American psychologist Edward C. Tolman created a form of behavioural testing suitable for a nocturnal and burrowing animal such as a rat or mouse: the maze. He showed that animals could learn their way around a maze without reward, and developed the concept of 'cognitive maps' to explain their behaviour.

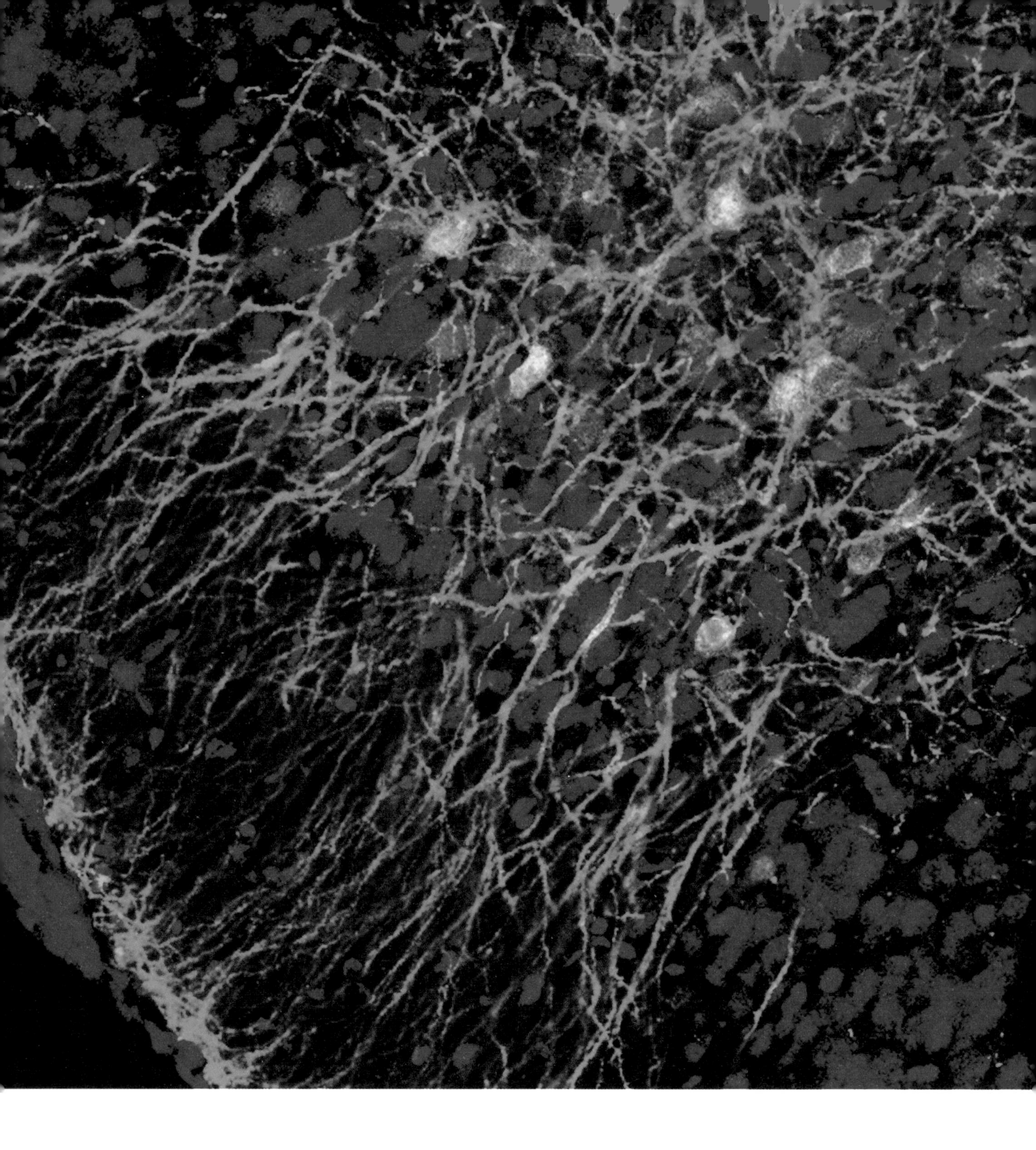

Fluorescent labelling highlights neurons that represent olfactory information in the mouse brain

the cells become active', says O'Keefe. The mouse runs on a ball in a virtual reality environment. 'We've had a lot of experience at the Institute of Cognitive Neuroscience in Queen Square using virtual reality to look at how human patients could navigate in different environments', says O'Keefe.

Space, it goes without saying, is the medium of the architect. 'My architecture starts in the spaces I create in my mind', says Ian Ritchie, the architect who has designed the SWC, and who heads an award-winning international architectural practice with many extraordinary spaces to its credit. They include the lift towers of Madrid's Centro de Arte Reina Sofia and the astounding 244-metre long arched hall for the New Leipzig Exhibition Centre. Both of these projects involved innovative engineering to achieve glass structures of lightness, transparency and deceptive resilience. Yet the full range of projects designed by Ian Ritchie Architects Ltd (IRAL) has no identifiable aesthetic, structural or material signature. Instead they are characterised by an intense analysis of the brief, followed by a solution informed by the building's purpose, the surrounding environment, and of course the budget.

Ritchie's design for the Courtyard Theatre, the 'temporary' 1000-seat performance space erected at breakneck speed for the Royal Shakespeare Company during the redevelopment of the Royal Shakespeare Theatre in Stratford-upon-Avon, is a rusty rectangular steel box that blends seamlessly with the colour of the surrounding brickwork. It came in at less than half the budget the board had previously been quoted for a smaller building. Ritchie had been nominated by the RIBA in 1997 to advise the RSC on the selection of an architect for its transformation project, and was subsequently appointed an RSC Governor. When the company's executive director Vikki Heywood suggested bringing forward the opening of a temporary theatre by a year, he offered the board a design that would, if all went well, be open for business a little over eighteen months later.

He threw himself into the challenge of creating a space that would work for actors and audiences, while meeting the potentially crippling requirements for low cost, high speed and the approval of the planners. The Courtyard was a resounding success, opening in time to host performances during artistic director Michael Boyd's Complete Works season of 2006–7, and demonstrating the exciting possibilities of the large thrust stage arrangement that was also to be introduced in the redeveloped Royal Shakespeare Theatre. Ritchie's artistic vision and enthusiasm for collaboration, allied with a grounded pragmatism that is not generally seen as a characteristic of high profile architects, won him many friends at the RSC including the RSC board's deputy chair, who also chaired the committee overseeing the transformation project in Stratford: Susie Sainsbury, the wife of David Sainsbury.

Susie Sainsbury recently rediscovered a sketchbook she had owned as a child, filled with floor plans for imaginary houses. Her professional career began not in architecture, but in educational publishing, first with Oxford University Press and then with Jonathan Cape. However, by the time she was entrusted with overseeing the transformation of a building freighted with the memories of generations of actors and audiences, she had had several opportunities to observe the design and construction of buildings at close hand.

During the late 1970s, Susie Sainsbury was the full-time mother of three small children and her husband the finance director of J Sainsbury plc. David's parents, Sir Robert and Lady Sainsbury (Bob and Lisa) had commissioned Foster Associates to build the Sainsbury Centre for the Visual Arts on the campus of the University of East Anglia in Norwich. Ian Ritchie was an architect with Foster's firm at the time, and worked with Norman Foster on its design concept between 1974 and 1976. The startling rectangular steel and glass building, said to resemble a hangar for an airship, houses the paintings, drawings and sculptures, ancient, modern and ethnographic, acquired by the couple in a long career of enthusiastic collecting and connoisseurship. It opened in 1978.

Less than a decade later David Sainsbury presented his father, as an 80th birthday present, with a model by Norman Foster showing how a possible future extension could be built to the Centre should the demand for space increase. 'Bob said "Wonderful, when do we start building?"', remembers Susie Sainsbury. The extension opened in 1991. A further £12.5 million refurbishment, including a new underground link between the original gallery and the extension, took place between 2004 and 2006. Susie was not directly involved in any of these projects, though she was an actively interested observer, and later took the lead in commissioning a book on the Centre by the leading architectural writer Professor Witold Rybczynski (*The Biography of a Building: How Robert Sainsbury and Norman Foster Built a Great Museum*, Thames & Hudson 2011).

It was a project on a much more domestic scale that gave Susie Sainsbury first-hand experience of the physical complexity on one

THE WORLD OF THE SENSES
Everything we know – or think we know – about the physical world around us comes through our senses: vision, hearing, touch, taste and smell. The brain's job is to turn these 'sense data' into meaningful information, by reference to previous experience, and to generate an appropriate response. If your brain detects that its visual representation of a car is rapidly getting bigger then you instantly know it is heading towards you and you had better get out of the way.

hand, and the human relationships on the other, that underlie the design and construction of a building. In the mid-1990s she and David restored a small country house in the Chilterns, originally dating from the early 18th century. 'That was where I learnt about construction and the importance of the working relationship between the design team and the client', she says. 'I was able to work closely with the architect and the local contractors. I know where all the pipes are under the floor, I made myself go up the scaffolding though I don't like heights, I learned all the important rules about how you build a building - and how to be a decent client, not changing your mind all the time.'

Armed with all these experiences, Susie was keen to be involved when her husband began to think about how to give his personal support to the scientific research that had so entranced him as an undergraduate.

MEMORY
Some behaviours, such as a baby's sucking reflex, appear to be inborn. But by adulthood we have added a vast repertoire of learned behaviours, from using a spoon to sending text messages. We also build up a core of knowledge about the world – that things fall down not up, that more distant things look smaller, that hard things hurt if you bump into them. Learning depends on changes in the strength of neural circuits.

As a very young man, David Sainsbury was handed a substantial share of the family business, anticipating his future career in its management. 'In the curious way my family does these things, I was sent along to see the family solicitor, Ethel Wix', he says. 'She said I should set up a charitable trust. Then she said, "What are you going to call it?" I said, "What do people call their trusts?" She said people call them after their favourite flower or favourite book. At that time my favourite book was *The Great Gatsby*, so I called it the Gatsby Charitable Foundation.' David's founding settlement for Gatsby consisted mostly of Sainsbury shares. At the time the company was still privately owned, and they made very little in the way of a cash return. In 1971 David went off to New York to do an MBA at Columbia University, and more or less forgot about Gatsby. Everything changed in 1973, when the company floated on the stock market as J Sainsbury plc. The flotation was hugely successful, the shares shot up in value, and suddenly Gatsby had serious money to spend.

David Sainsbury's first major venture into philanthropy established the Sainsbury Centre for Mental Health in south London, dedicated to exploring improvements to the delivery of psychiatric care after the closure of many long-term residential units. His next initiative arose from conversations with his old Cambridge friend Roger Freedman. Freedman had become fascinated by plant breeding, and had studied plant molecular genetics at the Plant Breeding Institute in Cambridge. 'Roger told me that genetic engineering would revolutionise plant research in a very interesting way', says Sainsbury. As he was interested in venture capital at the time he initially employed Freedman to investigate whether there were any commercial opportunities for investment. But after looking into the issue Freedman advised that it was too early to think about doing so.

Instead he observed that although we could now create new varieties of plant through genetic engineering, our knowledge of the most fundamental characteristics of plants – how they grow, how they fight disease – was woefully lacking. Botany had always been unfashionable, and had not been able to compete for funding with medically-oriented research in biology. With characteristic decisiveness, in the mid-1980s David Sainsbury handed him a budget to set up a plant sciences funding programme within the Gatsby Foundation. After this had been running for a short time, Freedman felt the time was right to make a more focused intervention. 'He said I should set up a centre for plant research', says Sainsbury. 'His other inspiration was that the focus of the research should be disease resistance in plants.'

The Sainsbury Laboratory for plant molecular pathology opened in Norwich in 1987 as a joint venture with the neighbouring John Innes Institute, the University of East Anglia and the Agricultural Research Council (now the Biotechnology and Biological Sciences Research Council). 'David believed it was important not to do things in the middle of nowhere, but as part of existing institutions', says Freedman. 'Then you get the right people, and set them free by giving them the resources they need.' Four bright young group leaders including David Baulcombe, who is now Sir David and Regius Professor of Botany at the University of Cambridge, quickly established the laboratory as one of the leading centres for plant disease resistance in the world.

As well as the satisfaction of backing a winner, David Sainsbury gained valuable insight into how best to support groundbreaking research. 'I took away the lesson that it's often technology that opens up a new area of science, not necessarily a scientific discovery', he says. 'It suddenly makes a whole area tractable – you don't know what it will produce, but it'll probably produce something interesting.' The second lesson he learned was that charities had a unique role to play in kick-starting research institutes that depend on such technological breakthroughs. 'While the peer-review system is best for funding research,' he says, 'it's probably not so good for situations where you have to make a big decision and take money from other people – who are the people sitting round the table making the decision. The research council system is intrinsically conservative. If money is pouring in it's much easier, but usually it isn't.'

The experience with the Norwich laboratory was highly influential in refining the philosophy of the Gatsby Foundation – as a funder that would take the lead on research projects that were setting out in new directions. With his term as chairman of J Sainsbury plc coming to a close, in the late 1990s David Sainsbury began to think about how Gatsby might support research in his first great interest: the science of the brain.

As with the Sainsbury Laboratory in Norwich, Roger Freedman was closely involved in helping David Sainsbury decide how best to support neuroscience through the Gatsby Foundation. At the time advances in technologies for investigating the brain had slowed: it did not look like a 'breakthrough' moment. Instead of an experimental laboratory, Freedman suggested a unit that would explore the potential of networks of brain cells through computer simulation. His researches had led him to exciting work at Oxford in the area of 'neuromorphic engineering' – a term coined by the CalTech engineer Carver Mead to describe a new way of designing computer chips with analog circuitry so that they worked more like networks of neurons. The research of Rodney Douglas, Kevan Martin and Misha Mahowald was underfunded, like much work in the field after the pessimistic 1973 Lighthill report into artificial intelligence research, and Gatsby made them a grant at a critical juncture. Later they left Oxford to establish the renowned Institute for Neuroinformatics in Zurich.

COMPUTATIONAL NEUROSCIENCE
Computers can be used to test theories about many aspects of brain function, such as learning, attention and perception. Computational neuroscientists model activity at every level, from the way information is encoded in trains of pulses in an individual neuron, to the dynamic interactions of different neurotransmitter systems underlying emotional responses. Separate but related research activity seeks to improve the capacity of machines to learn, based on our understanding of neuronal systems.

Meanwhile, wanting to know more, Freedman signed up for the three-week workshop in neuromorphic engineering held each year at Telluride in Colorado. There he met one of the chief architects of neural network theory, Geoffrey Hinton – an Englishman then working at University of Toronto – and invited him to head a new unit in London with funding from Gatsby. The Gatsby Computational Neuroscience Unit (GCNU) opened at University College London in 1998. Hinton was its first director; with him came another Englishman, Peter Dayan. 'I had been a post-doc of Geoff's', says Dayan, 'and was then an assistant professor at MIT. He asked me if I was interested in coming back to Britain to help found this unit. It was a great opportunity to work with Geoff to set this up.' In 2002 Hinton returned to Toronto, and Peter Dayan succeeded him as director of GCNU. Like

A computer algorithm grows crystal-shaped tree structures on top of a grainy image, which can then be computationally sharpened

the Norwich lab, the unit has established and retains an enviable international reputation in its field.

'Theoretical neuroscience delivers three sorts of analyses', says Dayan. 'Mathematical analysis provides theories bridging between different levels of explanation. It also provides data analysis: you could record from 1000 neurons, you might be interested in how closely correlated the activity of different neurons was. That's a statistical question. Likewise people are interested in recording an animal on camera as it goes around behaving. You have somehow to classify this data you get, which is an interesting computational vision problem. The third thing is that the brain is solving a lot of problems that we would like computers to be able to solve, such as vision, and so we need to have a way of understanding this. What's great about biology is that there's a reason why things are the way they are.'

From its inception, GCNU occupied the top two floors of a former office building in Queen Square, which UCL Provost Derek Roberts was able to rent on the strength of the Gatsby donation. At the same time the multidisciplinary Institute for Cognitive Neuroscience, previously a 'virtual' institute, moved in downstairs. Like many other buildings in which scientists work, it is a warren of rooms and corridors throughout which the accumulated clutter of decades of work has gently settled. As theoreticians, the GCNU's occupants have apparently simple needs: for individual spaces conducive to uninterrupted thought, a communal space in which to hold discussions, and plentiful opportunities to develop ideas through sketching, writing or calculating on whiteboards. Grand architecture, therefore, was not a consideration in the plans for the unit. It was another foray into plant sciences that inspired the creation of Gatsby's first landmark laboratory building. And as David Sainsbury frankly admits, his wife Susie played a critical role.

By the mid-2000s, the success of the Sainsbury Laboratory in Norwich encouraged Roger Freedman to propose another institute that would use the new tools of biology to understand the world of plants. This time the focus would be on how plants grow and develop, a vital question underlying efforts to improve the growth of agricultural crops to feed the world, as well as to promote biodiversity. Gatsby announced a partnership with Cambridge University in 2005, and the site chosen could not have been more sensitive: a strip of land along the north side of Cambridge's 19th-century Botanic Garden. Nevertheless, in the first instance neither Sainsbury nor Freedman saw the need for a high-quality building.

'I told David he didn't need a special building', says Freedman. 'Science changes – you shouldn't put a lot of effort into building something wonderful, because you're going to knock it down.' This argument made sense to David Sainsbury. But Susie Sainsbury, who was by this time experienced at collaborating with architects and contractors through her work for the RSC, disagreed. 'You have been given one of the most beautiful sites in the country', she told him. 'You must have a beautiful building.' She also made the point that if the building needed to be adaptable to take account of future scientific developments then that was something that should be put into the brief for the architect.

Once the Gatsby trustees had approved the funds for the new laboratory, Freedman visited a number of new labs around the world. He made a point of talking to the early-career researchers who shoulder the main burden of scientific production. 'When you create a new department and get some big shot to be the director, he designs his dream lab for the last time he was in the lab – which is 30 years out of date', he says. 'That's what I avoided by being ignorant and talking to people lower down.' He then drew up a list of three essential criteria for its design. The first was abundant use of natural light; the second was that the building 'should foster a sense of community and conviviality'; and the third was that its design should be adaptable enough to adapt to cope with unforeseen changes in the technology the scientists might want to use. He also invited Robert McGhee to act as laboratory design consultant on the project. McGhee is the architect who supervised the construction of the Howard Hughes Institute's Janelia Farm neuroscience and imaging research campus, to a design by Rafael Viñoly, on the Potomac River north west of Washington DC.

Meanwhile Susie Sainsbury ensured that a number of really interesting architectural practices, whose work she admired, were invited to apply. 'We told them, "We don't want your ideas of a finished building – we want a strategy for how you would work with the scientists and the client to provide a set of genuinely adaptable laboratories, and how you will use the historic and beautiful setting"', she says. Against the decided preference of Cambridge Building Services for a safe pair of hands who had previously designed a standard laboratory building for the University – a 'red brick box', as Susie Sainsbury saw it – the selection committee appointed one of the more adventurous architects, Stanton Williams.

Alan Stanton himself led the design team. Stanton, who had briefly been in partnership with Ian Ritchie at the end of the 1970s, went into partnership with Paul Williams in 1985, having previously worked on projects such as the

Pompidou Centre under Renzo Piano and Richard Rogers. Stanton Williams has produced a considerable body of work for museums and galleries in Britain. It had delivered the award-winning Millennium Seed Bank in Sussex in 2000 for the Royal Botanic Gardens.

The team produced a design for the Sainsbury Laboratory that sensitively integrated a modern structure into the surrounding landscape. The interiors are light and spacious, with views across the gardens, and planting in external courtyards that softens the transition from the building to the gardens beyond. Construction began in February 2008, was completed by December 2010, and HM The Queen formally opened the building in April 2011. It has delighted the scientists who work in it, both for its aesthetic quality and for the excellent research facilities, such as state-of-the-art growth rooms where plants are raised under strictly controlled conditions. Public recognition soon followed: the building took the RIBA's prestigious Stirling Prize in 2012 and won a number of other awards.

Susie had promised her husband that the project would come in on time and on or under budget. That it did so owed a great deal to others, including Stuart Johnson.

Stuart Johnson is a chartered surveyor and project manager who first came into contact with David and Susie Sainsbury when he supervised the refurbishment and new underground connection at the Sainsbury Centre for the Visual Arts. His management of that project and the introduction of a rolling programme of maintenance to avoid the need for major refurbishments in the future won their confidence. Johnson was appointed to manage the £82m Sainsbury Laboratory project.

'It became clear to me that like most clients, David Sainsbury and his charity needed early cost and time certainty', says Johnson. 'The procurement route I think works well for Gatsby is that you have the best possible team that you can, and enable them to do the best they can.' His approach sounds like common sense, but apparently it is not. 'What's usually done is that the architect does enough to get planning [permission], and the rest is done by a contractor', says Johnson. 'If the architect continues as contract administrator, he or she has the power to go on issuing architects' instructions all the way through the process. So the two or three per cent that should be their allowance for design development gets leveraged into 20 or 30 per cent. Costs overrun, and the whole thing gets delayed.'

This is exactly the nightmare scenario that David Sainsbury wanted to avoid in Cambridge. And Susie Sainsbury had not known whether Johnson would be comfortable working with Stanton Williams (though in fact he worked previously with high-profile architects, not least on the Sainsbury Centre for the Visual Arts). 'I had put everybody into a more risky environment', she says. She need not have worried. 'Alan and his team were just so good', she says. 'They committed to building what the client wanted, not what the architect wanted.'

Johnson reassured David Sainsbury that there weren't just the two alternatives, a cheap and ordinary building or an 'iconic' structure at risk of spiralling costs. His approach depends on careful advance planning, and early cooperation between the architect and the contractor. 'Before you start construction, you get the design team, led by the architect, to do a complete, coordinated design' he says. 'Then you shortlist contractors and find a preferred contractor to work alongside the architect, and comment on the project's buildability, the supply chain for different materials and so on. And then once everything is buttoned up, you sit down and negotiate with your preferred contractor a guaranteed maximum price. And you give them a premium for taking the post-contract design development risk. There are no fuzzy edges.' For the Cambridge plant science laboratory, this meant that once the contract with the contractor, Kier, was signed, Stanton Williams was 'novated' (in building contract parlance) to Kier during the construction phase. 'It means the contract is passed over from the client to the contractor', says Johnson, 'so that Stanton Williams worked under Kier's direction to finish the job.'

FEELINGS
The brain is the seat of emotion as well as reason. Feelings of joy, sorrow, anger, fear, love and disgust are all associated with particular brain regions and with changing levels of neurotransmitters. Emotional inputs to circuits involved in perception and memory help to direct our attention and attach significance to objects and events.

The plan worked in Cambridge to a degree that fully justified the trust David Sainsbury had placed in Johnson, whose own commitment to the ideals of the Gatsby Foundation is unswerving.

David Sainsbury works with a small, close-knit staff and team of advisers, who are notable for their loyalty and commitment to his vision. Since 2007 they have been headed by Peter Hesketh, who made the switch to philanthropy after 20 years in the defence industry and has never looked back. 'Here was this small, very agile, powerful little unit doing fantastic things', he says. 'It's a very collegial office, with no great hierarchy. Gatsby has a different rhythm and different operational style because we deliver projects, rather than being purely a grantmaker.' For the period of his time in office as Minister for Science and Innovation (1998-2006) in the Labour government led by Tony Blair, Sainsbury had been legally prevented from having anything to do with decision-making at Gatsby: it was up to the Trustees to award grants as best they could, following the themes he had already established. As soon as he left his post, Sainsbury actively engaged with Gatsby again, and invited Hesketh (who was already working for the Sainsbury Family Charitable Trusts) to become Gatsby's first chief executive. With Hesketh's help, he began to focus the work of the Foundation on riskier, longer-term projects. 'When I go to David and say, "This is impossible"', says Hesketh, 'he says to me, "Well if was easy, everybody would do it." That's his style. Let's take an issue, get under the skin of it, look at where we need to intervene and do that with resilience and determination.'

One such impossible task was to find a way of making a significant impact in the field of experimental neuroscience: through an institute with the deceptively simple but ambitious goal of finding out how the brain works. During his time in government, David Sainsbury had never stopped reading and thinking about the mysteries of the brain. At a dinner held to mark the 50th anniversary of the discovery of DNA in 2003, he sat next to the Cambridge molecular biologist Sydney Brenner. Brenner, who had won the Nobel Prize for Medicine the previous year for his work on the genetics, neuroscience and behaviour of the nematode worm *Caenorhabditis elegans*, had been a young don at King's College when Sainsbury was a student, and supervised Roger Freedman's PhD in molecular genetics at the Medical Research Council's Laboratory of Molecular Biology (LMB).

'Brenner was a great hero,' says Sainsbury, 'because he was a great scientist but also great fun. At this dinner I said, "I think neuroscience is becoming very interesting, who can tell me more about what's going on?" He mentioned Tom Jessell, a Brit at Columbia University in New York, and said, "If you ever want to do something in the UK, go and talk to him."' David Sainsbury had done his MBA at Columbia in the early 1970s, and retained strong links with the university. So the next time he visited he looked up Tom Jessell, now Claire Tow Professor in the Departments of Neuroscience and Biochemistry & Molecular Biophysics at Columbia. Jessell introduced him to Richard Axel and Eric Kandel, both Nobel-prizewinning professors at Columbia who had made extraordinary contributions to understanding neural circuits through studying simpler organisms such as flies and molluscs.

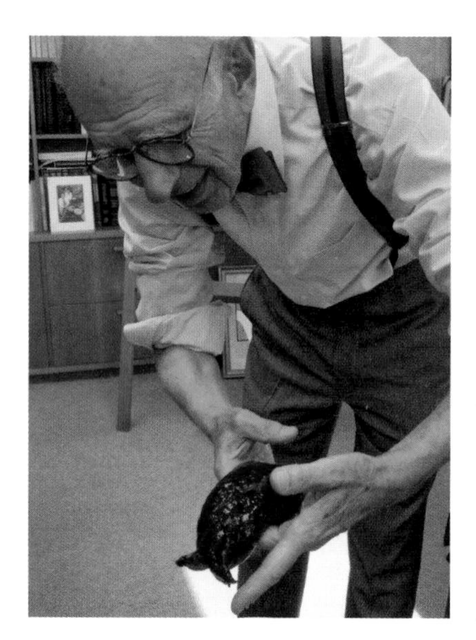

WIRING THE SEA SLUG
The American neuroscientist Eric Kandel chose to study the neural basis of learning in the sea slug *Aplysia*, which has a simple and accessible nervous system. Kandel traced the circuits involved in the reflex response of the animal to a light touch. He showed how repeating the same stimulus weakened the connections (habituation), or pairing them with a shock to the tail strengthened them (sensitisation).

'David visited me in my lab', says Axel, who later joined the Governing Council of the SWC, 'and I asked a graduate student to take him down to a two-photon microscope in which we were imaging a fly brain in response to two different odours. And you could determine the nature of the odour that the organism had encountered by the patterns of neural activity. And David spent a lot of time down there, seemingly fascinated.' Sainsbury says that the visit confirmed his view that it was an immensely exciting area. 'The research question that I had seen the beginning of – "How do neural circuits process behaviour?" – was exactly what they were asking.'

Once again he asked Roger Freedman to research the possibility of creating a new institute in this area. But this time Freedman said no. The plant sciences programme at Gatsby had expanded, the new Cambridge laboratory was already on the drawing board, and he was just setting up a US-based charity, the Two Blades Foundation, to develop disease-resistant plants. To take the neuroscience programme up another level at the same time he felt would be too much.

On leaving government in 2006 Sainsbury immediately went back to Columbia, where he was introduced to Sarah Caddick. Caddick is a British neuroscientist, who originally went to the US on a postdoctoral fellowship, but chose instead of a research career to take strategic roles in a number of philanthropic foundations. At the time she and David met, she headed the Centre for Neuroscience Initiatives at Columbia, and was deeply involved in the direction of a new Mind, Brain and Behavior Center (now the Zuckerman Institute), to be housed in a new interdisciplinary building on the university's Manhattanville campus. 'David asked me to come and help him set up an institute in London', says Caddick. Although she had not previously considered leaving New York, she accepted his offer and moved to London in 2007.

'It was very unusual for me to work with a philanthropist who had an understanding of both the science he wanted to invest in, and its context', she says. 'Having been science minister meant that he had been exposed to enough scientists to see what the back end of science looks like. What was remarkable was his willingness to work around all of the politics and the realities of academia.

Often if you have a lot of money to put into something, your expectation is that you can achieve almost anything.'

Before Caddick arrived, Gatsby's neuroscience programme had made a small number of grants to experimental laboratories, as well as a £10 million, ten-year programme to set up the GCNU. Her appointment coincided with a breakthrough moment in neuroscience, when new technologies were making it possible to explore neural circuitry to a previously unimagined degree. 'This is one of the most exciting times in terms of understanding the functioning of the brain', says Marc Tessier-Lavigne, President of The Rockefeller University and chair of the SWC's Governing Council from 2009 to 2014. 'We are beginning to understand the nature of the neural circuits that underlie higher order functions including perception, cognition and of course behaviour – all of which are linked. What's made that possible is the convergence of many technologies – the tools that allow you to trace the connections between neurons have improved dramatically. But to really get at causality you have to be able to intervene: you have to alter the firing pattern, and again the technologies that make that possible have now come of age. So now it's off to the races. The time is right now to focus on understanding the function of neural circuits as they underlie complex behaviour.'

WHAT IS A CIRCUIT?
At its simplest, a circuit is a set of connected neurons that collectively process a given piece of information, such as the reflex that makes you pull your hand away from a hot saucepan. In a snail it may take only a few neurons to control a simple withdrawal reflex. In mammals circuits may involve hundreds or thousands of neurons.

Partly based on his experience with the Norwich lab, David Sainsbury realised that this was a strong argument for setting up an institute, rather than continuing only with a distributed grants programme. 'The rationalisation was that where you have these breakthrough technologies, there is merit in having them together in a new institute', he says. 'Being able to exchange technology, interacting, is the way to do it. Forty or fifty years ago you couldn't have done this: it was too early.' Needless to say, he had continued his conversations with Richard Axel and others at Columbia as he developed his idea.

'Initially it was an idea of forming an institute that freed the investigators of the worry of support, so that they could address interesting problems in neuroscience in a highly interactive way', says Axel. 'Gatsby had already supported a small group of theoretical neuroscientists, and David felt that the liaison between theoretical and experimental scientists was important. Too frequently the experimental neuroscientists could not speak the language of the theoretical neuroscientists, and the theorists cared too much about the beauty of the theory and not enough about the correlation of the theory to the real biological world. So the idea is to constrain the theoretical thinking by dirty facts.'

'David had spoken to a number of scientists in New York and other places', says Caddick. 'He had an advanced idea of the goals of the institute – to understand how the brain generates behaviour – and he knew that it could only be a good thing to integrate the theory and experimental science. I was charged with figuring out the optimal way to make that happen within the parameters that he'd set.' David had an approximate budget and a timeframe – he wanted to see the institute open sooner rather than later. Caddick's job was to take those parameters and come up with a workable proposal, both for the size of the building and its staff, and the kind of science that might be done within it.

'Neural circuits and behaviour sounds nice and contained,' she says, 'but it's not. It's trying to pull apart what we mean by a circuit. Is a circuit simply the minimum set of bits and pieces that's required to deliver a particular behaviour, whether that's reaching for a cup of tea or determining that that lion is going to kill you if you don't run very quickly? Even that one question alone could take up an entire university of neuroscience researchers.' Given the more realistic parameters of the projected institute, Caddick proposed that Gatsby take a relaxed view of exactly which research questions it might address, and the techniques it might deploy. 'It isn't rocket science' she says. 'You get smart people that are exploring anywhere along that continuum and ultimately you will advance knowledge.'

One parameter was not negotiable: that the location of the institute would be in the UK. 'I have always supported basic science projects', says David Sainsbury, 'because I think that is where the really important breakthroughs are made. But equally I have always believed that where breakthroughs are made that can bring major economic or social benefits, there should be systems in place to ensure that these benefits are realised. I wanted the benefits of research at the Sainsbury Wellcome Centre to accrue in the first place to British industry and society.'

With her wide network of contacts in the area, Caddick soon became aware that the Wellcome Trust, a charitable foundation and the UK's biggest funding body for biomedical science, was also thinking about a major project in neuroscience. She quickly decided that cooperation would be a better strategy than competition, and so planning for the project acquired a further dimension.

TRANSGENICS

By deleting or inserting DNA sequences in the genome of a mouse embryo, neuroscientists can develop strains of mice with particular characteristics and study their neural circuitry. For instance, Jeff Lichtman and Joshua Sanes at the Harvard Brain Center have created transgenic mice in which different types of brain cell are tagged with different coloured fluorescent proteins. Their 'Brainbow' project hopes to trace the complete wiring diagram of the mouse brain.

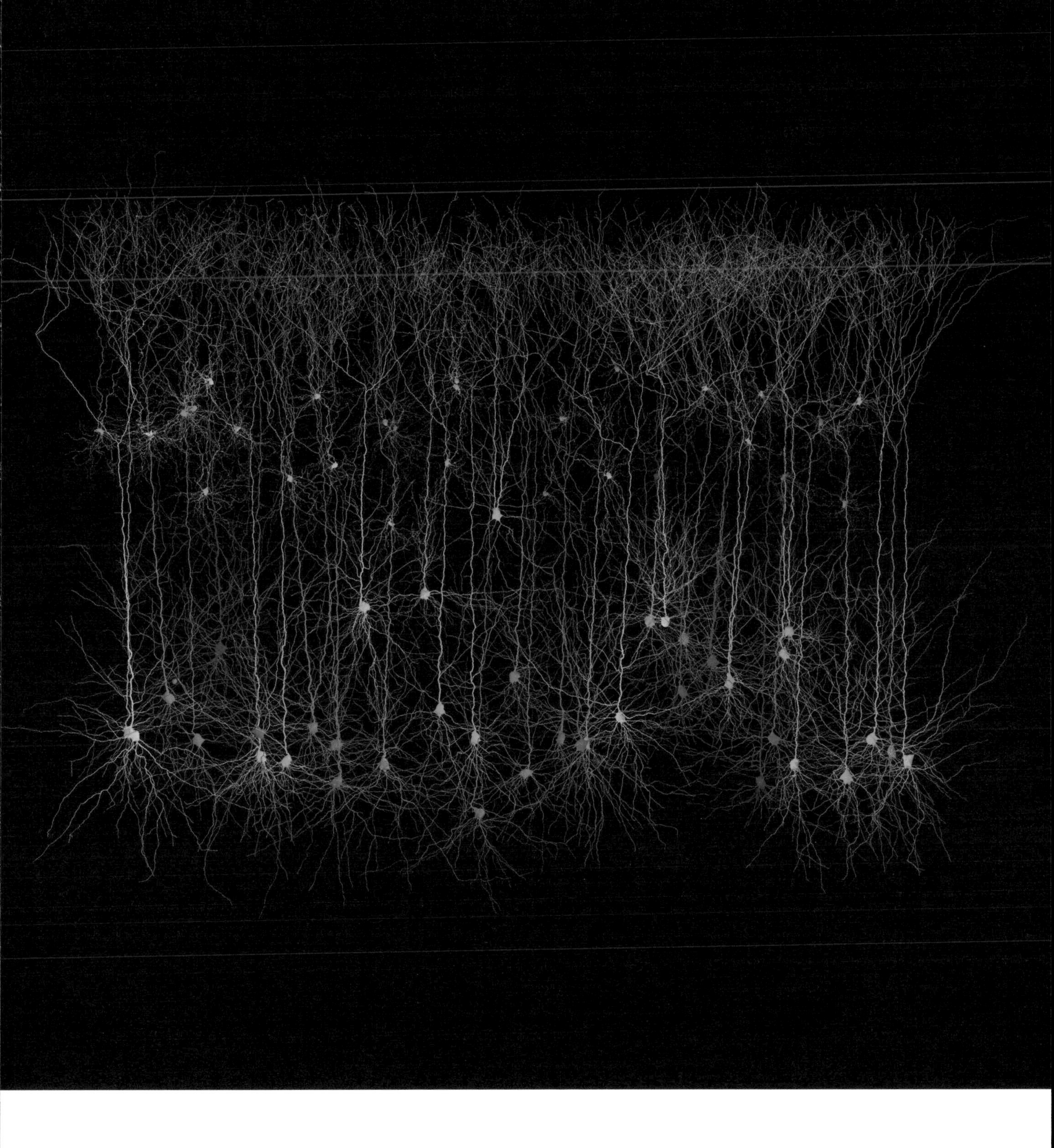

Synthetic cells in the neocortex created using
software that mimics optimal network formation
in real brains

Richard Morris is Director of the Centre for Cognitive and Neural Systems at the University of Edinburgh. His research focuses on the neurobiology of learning and memory, also one of the themes of the Sainsbury Wellcome Centre. He pioneered a now widely-used technique for testing the spatial memory of freely swimming laboratory rats, called the water maze, and established the essential role of a class of neurotransmitter receptors, called NMDA receptors, in encoding new memories. In 2007 the Chief Executive of the Wellcome Trust, Mark Walport, told him that the Trust was planning to make a £100 million investment in neuroscience, and offered him a three-year secondment to advise the Trust on how the money should be spent.

'I told him that from the point of view of science, there was a very obvious thing to do', says Morris. 'The Wellcome Trust has a fantastic reputation at one end of the scale of science – genetics, with the Sanger Institute and the sequencing of the genome – and also a terrific reputation at the other end of the scale, with the Wellcome Trust Centre for Neuroimaging in London, which is a pioneer of non-invasive imaging in human brains. But what's missing is the middle of the scale: how the cells and circuits actually control behaviour.' Walport and the Wellcome governors were enthusiastic about focusing on this level. Morris started his new position in the autumn of 2007, and discussions were under way about an institute. 'Until about ten years ago that middle level was ignored – the necessary technology was simply not available', says Morris. 'Things like the electrode giving way to the microscope as a way to do physiology, and the opportunity to use all sorts of genetic tools as interventions rather than just messy drugs or lesions like the psychologists were using, were great advances.'

MICROSCOPY IN ACTION
New optical microscope technology can reveal functional connections within a sample of brain tissue. Lasers scan through a fluorescently-labelled sample, taking optical slices and building up a 3-D image. Two-photon microscopes use light of such low energy that it takes two photons to excite one molecule of fluorescent dye. This produces an even sharper image, enabling scientists to watch the chemical interactions between live cells in vivid colours.

At this point Morris knew nothing of Gatsby and had never met Sarah Caddick, who had been on the Gatsby staff for only a few months. 'I told David Sainsbury that I was thinking of going to see Wellcome, to tell them what we're doing, and to suggest that we look at partnering on this', says Caddick. 'The UK cannot sustain two institutes of this size. First of all, if neither institute is fully funded they are going to suck up the lion's share of neuroscience funding from the research councils, and that is not going to make the rest of the UK happy. Secondly, the market for recruitment is going to be extremely competitive. And thirdly, right now Gatsby is not a household name. People are most likely to trust Wellcome because of the brand name.'

With Sainsbury's permission, she approached Morris. 'We then had a series of meetings and built up confidence and by the end of October decided we could work together', says Morris. 'So I then wrote a paper which I presented to the Governors of the Wellcome Trust at the end of the week before Christmas. I asked for £50 million and they said yes. The die was cast as quickly as that.'

Sir William Castell, the Chairman of the Wellcome Trust, knew David Sainsbury and was an enthusiastic supporter. Morris then met Peter Hesketh and started to work with him on how to apportion the total funds – which with Gatsby's contribution totalled £140 million – between the building, the staff and the running costs. 'It was a great thing for Sainsbury and Wellcome to come together and really make a significant impact', says Morris. The agreement between the two charities commits them to enduring support for core costs and maintenance. 'Wellcome will christen this a Wellcome Centre of Excellence, so that on application and approval there will be an expectation that they will continue funding, and we will do the same', says Hesketh. 'The first year will be a big programme because we put in cash for fellowships and so on. The budgets were set around 2008–9, and when we went out for recruitment, some of the players said "You need to be in tune with what the market is spending at the moment." So David put more on the table, and Wellcome did the same. It is a real partnership in that sense.' The total budget for the project, including the building, staffing and running costs for five years, eventually came to £150m.

Wellcome's Project Director, Dave Scott, became the Trust's non-executive representative on the SWC project. He brought with him a wealth of experience on complex laboratory buildings. Having first joined the Trust in 2002 to work on its flagship genetics centre, the Sanger Institute near Cambridge, he subsequently spent 6 years developing plans for the vast Francis Crick Institute near Euston Station in north London, a biomedical research centre jointly developed by Wellcome, Cancer Research UK, the Medical Research Council and three of London's leading research universities. 'These were much bigger projects than the SWC', he says. 'But they involve the same design issues, the same risks – just different scales.' First the two organisations had to decide which academic establishment should host the institute. By the spring of 2008, when the Wellcome trustees formally made the decision to fund the institute, Caddick had already begun to narrow down the range of options. 'I looked at every university across the UK', she says, 'and the reality is that for what we want to do and the timeframe we want to do it, there really are only three places that could conceivably host, nurture and bring it to fruition. It's the golden triangle: Oxford, Cambridge and University College London.'

Wellcome accepted the shortlist, with the proviso that an international panel be appointed to make the final choice. Wellcome nominated Sten Grillner, from the Karolinska Institute in Sweden, a former chair of the committee that selects the Nobel prizewinners in physiology or medicine. Gatsby put forward Richard Axel, as well as Gerry Fischbach, former Dean of the School of Medicine at Columbia. Sir Paul Nurse, then President of Rockefeller University (and currently Director of the Francis Crick Institute in London) reviewed the proposals on an ad hoc basis. With Sarah Caddick and Richard Morris, this was the group that scrutinised the bids and interviewed the applicants.

'We produced a bid that was rather generic', says John O'Keefe. 'We said we wanted to investigate how activity in neural circuits represents ideas, thoughts, actions and so on. We did not specify what part of the brain we were going to look at, who was going to be involved, or who would lead it. It was clear that a lot of the questions were now beginning to look like they were addressable within 20 years. The idea that you would be able to record from large numbers of cells in freely moving animals doing things which were of interest; that you now had ways of controlling cells, so that you were not only looking to see what patterns were expressed but you could impose those patterns: none of that was doable when we made the bid, but we could see that in a short time techniques such as optogenetics, two-photon microscopy and silicon probes would let us do those sorts of things.'

UCL put together a small group to prepare the bid, including professor of neural computation Michael Häusser, professor of cognitive neuroscience Jon Driver (now sadly deceased), Peter Dayan, and O'Keefe himself. Everyone acknowledges that Jon Driver, who took his own life after a motorcycle accident left him with unbearable chronic pain, was the central figure. 'He really was the driving force behind UCL's application', says Caddick, 'curating a series of visits for me and Richard Morris, and ensuring everyone was on the same page. I think without his efforts the UCL application may have fallen far short of where it needed to be.' The team also benefited from direct support from the Provost, Malcolm Grant, and the senior management team. 'I was hugely excited about the possibility and

CONSCIOUSNESS

Humans may be unique in having a self-awareness that extends to a detailed memory of the past and a sense of our continued future existence. Our social lives depend on accurately interpreting the mental states of others. The neural basis of consciousness remains elusive: some argue that it is an 'emergent property' of the collective activity of neurons in the brain, just as the patterns made by flocks of starlings in flight are emergent properties of the actions of many individuals.

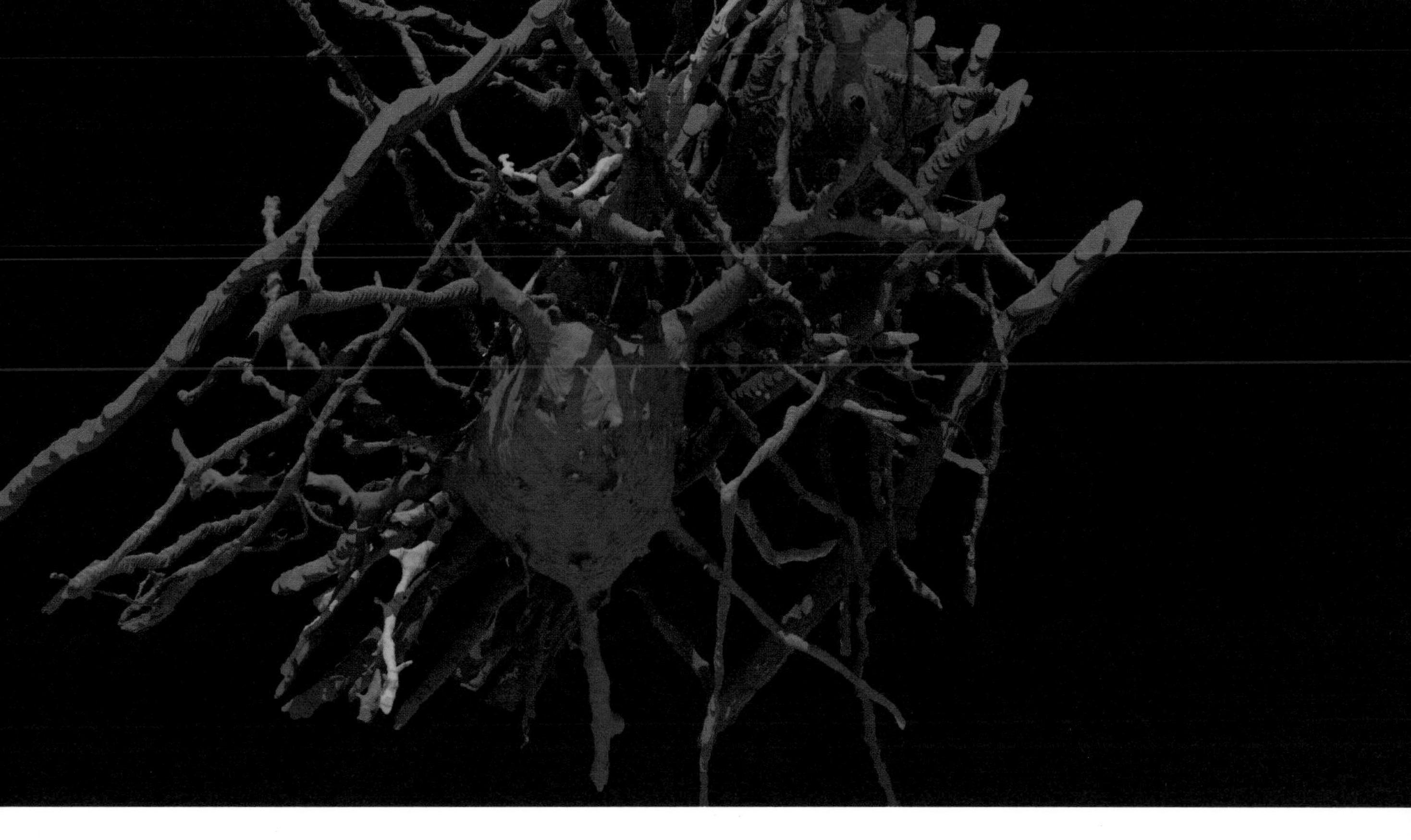

determined to win it', says Grant (now Sir Malcolm, and the Chairman of NHS England). 'In the whole area of neural circuits we already had great expertise. We knew there was stuff that we were doing that was at the cutting edge globally, and the Centre would be a good platform for taking that forward. At Gatsby's suggestion we proposed that the Gatsby Computational Neuroscience Unit become part of it. That was something the others could not offer.'

'We took advantage of the fact that there were huge numbers of neuroscientists at UCL', says O'Keefe. 'Sarah Caddick asked us "How many neuroscientist PIs [principal investigators] are there at UCL?" I said 100. It turned out when we did the numbers there were more like 450.' Caddick had asked each of the three universities the same question. 'At UCL I wrote a number on a piece of paper and gave it to Jon Driver, and said, "Don't open that until you have done your audit"', she says. She was only out by about 20. 'The bid acted to nucleate the neuroscientists at UCL who had been a bit amorphous before', says Dayan. 'There was a very intensive process within the university, driven by the vice-provost Ed Byrne.'

The process of preparing the bid went on for months, and was arduous for the scientists involved. 'We had a lot of meetings and presentations of the document

Connectomics: generating detailed maps of the connections between the cells and the regions of the brain

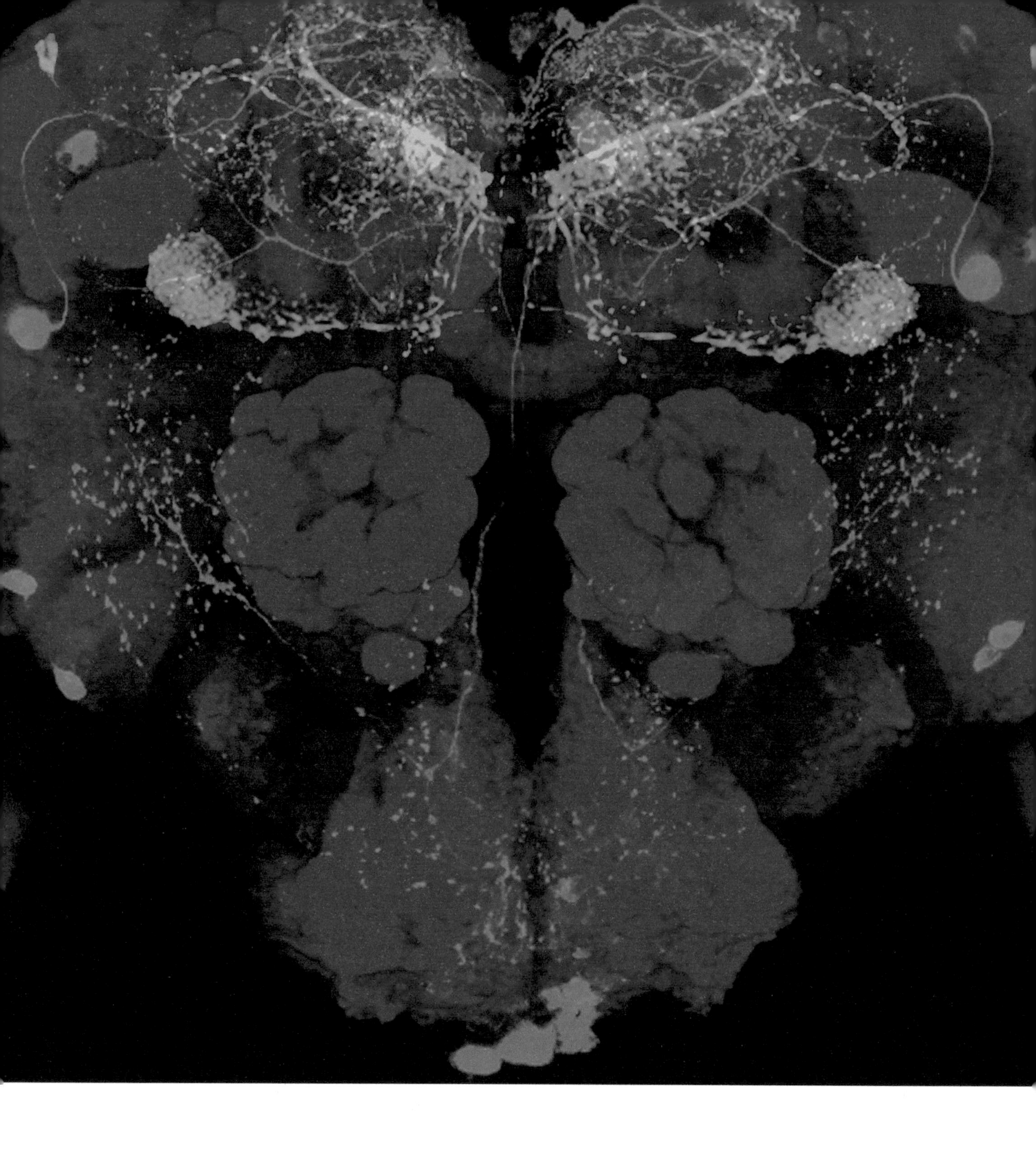

Neural circuit providing motivational control
in the fruit fly

and the bid', says O'Keefe. 'Like we do for younger people who go for fellowships – we put them through a whole bunch of mocks. But we knew what we were going to say, and there was no internal fallout. Even if we had not got the centre it would have been a positive experience.' The Provost, Malcolm Grant, took a personal interest in this process. 'I chaired a succession of meetings of the team and other senior staff to assure myself we were getting our act together', he says. 'There were then seven or eight rehearsals, to each of which the team invited a different group of senior scientists, who were asked to tear the presentation apart. So the proposition had been very carefully tested before it went into the final presentation. I was satisfied that the team was as strong as it could be.'

Their preparation paid off. 'On the basis of the breadth of the faculty and the insight of the presentation, we chose UCL', says Richard Axel. Having Peter Dayan's Gatsby Unit already at UCL was a key factor. 'Our bid was designed to have a very close interaction between theory and experiment', says Dayan, 'which turned out to be very attractive to the funders.' Needless to say the UCL team was euphoric. 'I was sitting in a restaurant in rural New Zealand with very poor connectivity', says Grant, 'when David Sainsbury rang up to tell me that we had succeeded. It was absolutely wonderful.'

The runner-up was Oxford, but the ancient university did not go away empty-handed. Its bid had been focused around the work of the Austrian scientist Gero Miesenböck, Waynflete Professor of Physiology and the originator of the technique of optogenetics. 'Oxford University put a significant amount of money on the table as part of the bid', says Sarah Caddick. 'We then made a smaller gift [£5m] and they set up the Oxford Centre for Neural Circuits and Behaviour under Gero's direction.' The Wellcome Trust matched Gatsby's gift, and Oxford's Vice-Chancellor, John Hood, agreed that the money originally offered by the University would be spent on the new centre. 'This exemplifies David's general view that if you're focused and interested and passionate about something, you can encourage other people to up their game', says Caddick.

The agreement between Gatsby, the Wellcome Trust and UCL was formally signed in September 2009. By that time a new phase had opened in the planning of the new institute: the choice of an architect to create a building that in Sarah Caddick's words would be 'creative, useful, adaptable and wonderful'.

OPTOGENETICS

Optogenetics enables experimenters to turn neuronal activity on or off on a timescale of milliseconds, simply by shining light on nervous tissue. It involves inserting genes that encode light-sensitive channels or enzymes that exist naturally in bacteria and algae into the genetic material of other species, including worms, flies and rodents. The light-sensitive molecules can be targeted to influence the activity of specific classes of neuron to reveal their role in neural circuits.

2

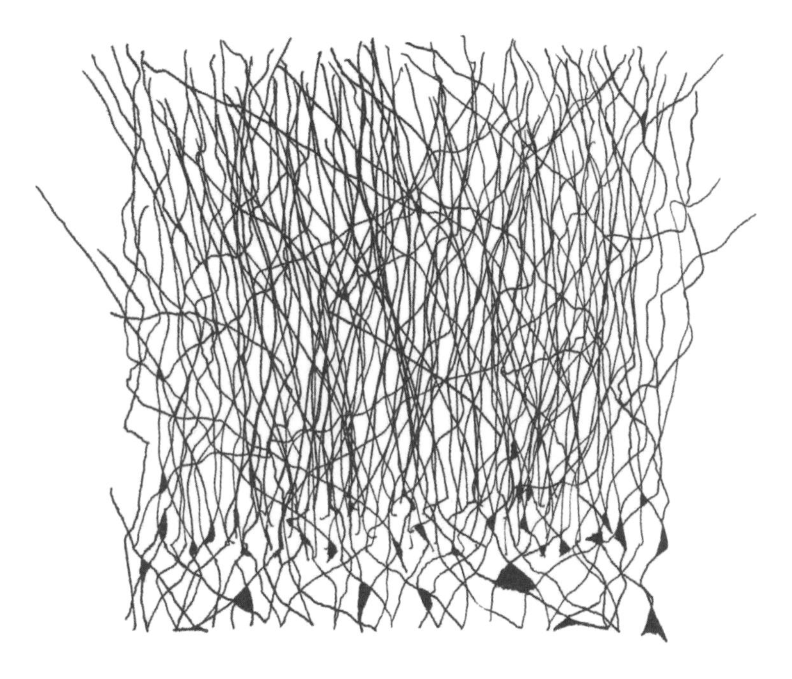

'My architecture starts in the spaces I create in my mind. Space is in here and out there, it is a continuum between inside and outside, mental and physical... Now I am designing with the mind in mind.'

Extract from 'Dreaming of a project', from *Lines* by Ian Ritchie, 2010

THE SPACE BETWEEN

LEARNING ABOUT LABORATORIES

Ian Ritchie Architects Ltd (IRAL) is a close-knit practice based in a converted 19th-century building in the Limehouse area of London's Docklands, formerly the offices of a shipping company: close to Canary Wharf, it has excellent land, water and air transport links. They have since more than doubled the size of the building with a stainless steel and glass extension. Ritchie set up the practice in 1981, and several of his colleagues have been with him for more than two decades. The practice operates collectively as a studio with a set of common ethical and design standards, based on a detailed understanding of the client's requirements. 'It's not our building – it's the client's building', says Chris Russell, a senior project architect who joined the practice in 2001. IRAL's interests embrace sculpture and engineering as well as architecture: perhaps the Spire in Dublin, a stainless steel cone rising to 120 metres above O'Connell Street from a base only three metres across, best represents all three. And, as they had shown with their design for the RSC's Courtyard Theatre, they enjoy a challenge.

When Stuart Johnson posted the invitation for architects to tender for the building of the Sainsbury Wellcome Centre for Neural Circuits and Behaviour, he received well over 100 initial applications. That was not surprising – the procurement process involved a Europe-wide invitation to tender, and anyone could apply.

'When I met the Provost, Malcolm Grant, with Peter Hesketh, they accepted that the project would be driven by Gatsby', says Johnson, 'and that I would be the driver as the funders' representative.' While in most contexts the title of such a role would be 'strategic project manager', UCL convention uses the term 'project sponsor', and that is what Stuart Johnson became. UCL had also asked that in recruiting the architects and contractors, they used the OJEU procurement process. OJEU is the *Official Journal of the European Union*: OJEU procurement is shorthand for the EU-wide system designed to combat corruption in the placing of public sector contracts. Johnson was happy to comply.

'I told them that there should be a phased approach to selecting the team', he says. 'I wanted a three-stage appointment process. With OJEU, you can't control who can submit. I said that we needed a first stage that would allow us to see the wood for the trees and take out the 60–80 firms that didn't have the right credentials. Stage 2 was about how the candidates would take ownership of the initial brief, what their process would be to turn that into a building brief,

LABORATORIES 1: THE FIRST LABORATORY?

The English Renaissance polymath and statesman Francis Bacon described in his unfinished fable *New Atlantis* (1627) a college called Salomon's House. Its purpose was 'the knowledge of Causes, and secret motions of things... to the effecting of all things possible.' The House was equipped with all possible forms of controlled environment, instruments and observatories. It inspired the foundation of Britain's first scientific academy, the Royal Society, in 1660.

The Courtyard Theatre, designed for
the Royal Shakespeare Company by
Ian Ritchie Architects Ltd and opened in 2006

and an indication of their fee. Then we would go to a more intense and detailed phase, with costings, for stage 3.'

The first step towards establishing the brief took place at a meeting on 2 December 2008, before the formal agreement between UCL and the two funders had even been signed. Decisions taken at this meeting, augmented and modified through subsequent advice and comment, formed the core of a briefing document on the project, under the codename 'Glimmer', completed by mid-July 2009. It began with a clear focus on the research direction to be taken by the institute's eventual occupiers:

> *The scientific remit of Project Glimmer is* Neural Circuits and Behaviour. *Research will be conducted using state-of-the-art molecular and cellular biology, imaging, electrophysiology and behavioural techniques, supported by relevant computational modelling and imaging resources, to identify the functional and anatomical determinants of information processing in the brain. Facilities are likely to include novel types of infrastructure and support (for example an advanced machine workshop) to enable the Centre to stay rapidly abreast of an exciting and fast moving field. It is anticipated that the Centre will ultimately comprise up to 12 experimental (wet lab) research groups and the existing Computational Neuroscience Unit (5 faculty, approximately 30 students and postdocs and 3 administrative staff).*

Nothing in this outline has since been contradicted. As for the design of the building, the brief could not have been less prescriptive. Rather than setting out what the building should be like, it focused on what the building had to do. It was specific about the space requirements, both in total and in relation to different types of accommodation (wet labs, write-up areas, seminar rooms, café etc), about the demanding level of service provision to be balanced against energy efficiency, and about the need for short- and long-term adaptibility. But as for the internal and external appearance, Johnson summed it up in a sentence: 'The building needs to convey a sense of quality and ambition while remaining prudent and cost effective.'

'David [Sainsbury] said, "This needs to be about science and function, and not a big iconic building"', says Sarah Caddick. 'We wanted an architect who would agree to spend the first couple of

LABORATORIES 2: SCALING UP SCIENCE

The French chemist Antoine Lavoisier revolutionised chemistry with his method of separating water into oxygen and hydrogen. He also promoted the idea of bespoke laboratory spaces, able to accommodate a wide variety of instruments and provide appropriate environments for different types of experiment. His own laboratory at the Petit Arsenal, where he moved in 1775, had an inventory of more than 8000 vessels, instruments and machines devoted to research.

years working with the scientists to design it from the inside out. And the whole project team had to be committed to the same approach. Gatsby had a precedent in the way it had approached the Sainsbury Lab. So we said, "Let's adopt that model". But this is a very different type of research building, and on a very different kind of site.'

Those architectural practices that made it through the team's initial sieve were invited to interrogate the brief, comment on its feasibility, and put forward a plan for how they would go about designing such a building. The selection process brought the number of candidates down to six, and eventually three. The day of the final architects' interviews was highly charged, as Susie Sainsbury remembers.

'The selection panel was much bigger than we had had for the plant sciences laboratory in Cambridge', she says. As well as Susie Sainsbury herself, Richard Morris, Sarah Caddick, Stuart Johnson and Peter Dayan, there were Dave Scott, Project Director for the Wellcome Trust, and Dave Smith, Deputy Director of UCL Estates. Sir Duncan Michael, former chairman of Ove Arup and chair of the board of the UCL hospital group, was there as an independent observer. 'The panel was large,' says Susie Sainsbury, 'but there was no one we could have left out.' Just as the selectors had done in Cambridge, they asked the candidates how they would approach designing for the scientists and the rapidly evolving science, the difficult central London site, the complex security and servicing. The competition brief was clear that, unusually for an architectural competition, it was looking for the capacity to understand that a conventional laboratory would not be the answer.

One of those who made it to the final stage was IRAL. Ritchie's previous work did not include any laboratories, although early in his career he had worked on the Cité de Sciences at La Villette in Paris, a pioneering centre for public engagement with science. So why did he put in a bid? Undoubtedly he was encouraged by the success of the Courtyard for the RSC and his new underground station at Wood Lane. But his interest in the work of the laboratory went far deeper than these.

'The whole area of neuroscience is fascinating to me', he says. 'There's also something fascinating about the link with ideas about how we conceive space, although at the time I was totally unaware of John O'Keefe's work in that area. But we got down to the interview because we believed that the fundamental question was, "Present a

LABORATORIES 3: UNITED AGAINST DISEASE
Lauded for his contributions to the germ theory of disease and his development of a vaccine against rabies, the French microbiologist Louis Pasteur founded the Pasteur Institute in Paris in 1887. He chose an interdisciplinary team of group leaders to work on infectious disease and teach microbiology. The Institute has since made numerous important medical discoveries, including vaccines against TB and polio, and the two forms of HIV.

strategy to achieve a new and correct and adaptable environment for neuroscience research, delivered on a tight site in Central London." So we wrote a strategy – we didn't design a building.'

For Richard Morris, who came to the meeting with a prejudice in favour of a more concrete proposal, the experience of the interviews was eye-opening. 'One company came along with a model', he says. 'And Susie said "You haven't talked to the scientists yet – how do you know what the building will look like?" There was a pregnant pause.' Another firm showed a design that included furniture layouts and their own, very recognisable, signature roofline. And then it was Ian Ritchie's turn. 'Ian comes along and says, "I haven't a clue what the building will look like"', says Morris. 'He said his plan would be to take his team and visit some labs, and find out about good practice and bad practice. He then gave examples of where he'd done this previously, such as Bermondsey Underground station, and how you go through the process of informing the architect of your needs, and how the

Virtual reality environment for experiments on spatial orientation at UCL

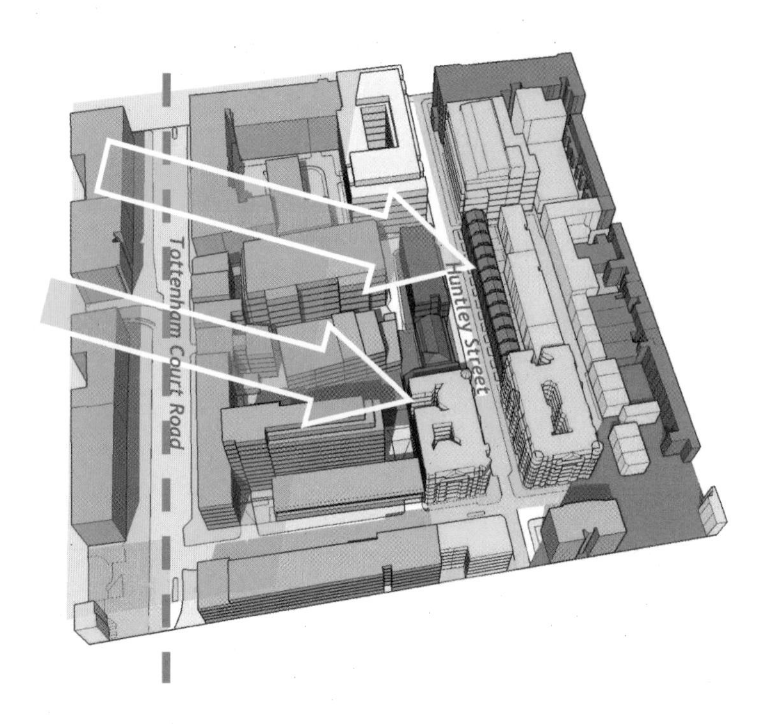

architect helps the client to jump out of their comfort zone into something better. It was a revelation.'

Ritchie displayed a confidence in his colleagues and an understanding of his prospective clients' needs that other candidates had lacked. He shared the presentation with Gordon Talbot, who would act as project director, and Chris Russell as senior project architect. They brought life-sized photographs of others who had to be left off the presenting team but would work on the building. Russell remembers the intense engagement between the panel and the presenters: 'There was a lot of discussion.' The IRAL team had worked hard to prepare a presentation that had no words on the slides, only images – images not necessarily associated with either architecture or neuroscience. 'One was of six people on a bicycle,' remembers Russell. 'One of the things that we stressed was the continuity of the team and the people,' says Ritchie. 'And we said that the three people they were looking at would be there at the end.' They were.

He also took something of a gamble (although it was a key part of 'interrogating the brief') by casting doubt on the suitability of the proposed site on Huntley Street, very close to UCL's main campus on Gower Street. 'We indicated that there might be an issue with rights of light on the given site', he says. 'I debated whether we

Architects' diagram illustrating the issue with rights to light on the proposed Huntley Street site

should suggest that. But our initial assessment was that it could be tricky.' The other risk he took was to ask for time to conduct real research. 'We emphasised the fact that it was impossible to design a building until we knew a lot more about the current and emerging techniques in use in neuroscience, their use of space, and the behavioural patterns of neuroscientists and their support staff', he says.

This was music to the ears of the majority of the panel members. 'Most were excited by the originality of Ian's thinking and approach', says Susie Sainsbury. 'Stuart was brilliant in the summing up at the end. There was already a majority in Ian's favour, and by the close of proceedings the vote was unanimous.' Ian Ritchie Architects Ltd was duly appointed to design 'Project Glimmer', now formally known as the Sainsbury Wellcome Centre for Neural Circuits and Behaviour.

Within weeks another team came on board, as essential to the process of creating a building as the architects: services and structural engineers. Crucially, the client invited the architect to be part of the selection panel. Through the OJEU process, submissions were filtered and following the final interviews, Arup, a global engineering company, won. Entirely coincidentally the SWC would ultimately be located opposite Arup's HQ on the opposite side of Howland Street, not the original location put forward by UCL.

Jennifer DiMambro is from Arup MEP (mechanical, electrical and public health) – which she defines as 'all the stuff that goes inside a building and makes it habitable or comfortable'. She appreciated that there would need to be a somewhat drawn-out process for developing the building before the contractor came on to the site.

'I do a lot of lab-type projects', she says. 'The focus is about enabling the science. Whereas with office buildings or theatres, it's more about the comfort and experience of the building.' DiMambro had led the Arup team on the Sainsbury Laboratory in Cambridge, which had many of the same people as the Glimmer project, including Stuart Johnson as project manager. 'The only real difference was the different architect and different approach', she says. 'For both projects you don't have the true end users, but you have user representatives. You can't say "I'm building this building and in four years time there'll be a job for you." You have to build the building and then right at the end say "Do you want to come and work here?" You have to second-guess what people want. That's pretty typical in lab design these

LABORATORIES 4: A FACTORY FOR IDEAS

In 1925 4000 engineers and scientists working for American Telephone and Telegraph (AT&T) and Western Telegraph were brought together in New York as Bell Laboratories. Their brief was to work together to solve fundamental problems that might have some future impact on the telecommunications industry. Discoveries including the transistor, the laser, and information theory have since been made at Bell Labs, and its research has garnered eight Nobel prizes.

days. The whole ethos has changed from being highly bespoke to being adaptable, so that the building has more longevity. Science changes, the focus changes, whether it's commercial or academic research. Having this more adaptable building typology is what the philosophy has been for the past ten years or so.'

Joining the team from Arup Structural Engineering was Christian Allison – his job is to make sure not only that the building stands up, but that the structure performs as required in terms of noise and vibration. 'It's really the reasoning behind the project rather than the economic drivers that interests me', he says. 'It's about what the building can deliver. At the start of Glimmer, the interesting thing for Ian and the client was that they were trying to develop a building that not only enables the science but also enables scientists to interact. What I'm most interested in is to see how that interaction will work in the finished building. One of the brave decisions that the client made is that they chose Ian, who had never done a lab before but put a lot of focus on how a lab may work, rather than an architect who'd done 10 labs and had a fixed idea of what a lab was like.'

The design team met for the first time at IRAL's office on 22 October 2009. Ritchie asked each member of the team to come to the meeting with something that illustrated their approach: he himself read one of his poems. Ice broken, the engineers immediately began to work with IRAL to understand the requirements of the building and to evaluate the site in Huntley Street that was giving Ritchie such cause for concern. Through weekly meetings that would continue up to and beyond the point at which construction began, they established a relationship that would be crucial to the success of the project. 'An architect's output is their drawing and their visuals', says DiMambro. 'We produce drawings at the end of the day, but our main output is all the maths, making sure that the building stands up, that you've got the right temperature, that you've got enough air. And it's not just a matter of them handing over a drawing and us going to make it work, it's an iterative process. The architect produces the idea of where we are going, and we think about how that can be realised.'

Most of what engineers do is not readily obvious to users of the building – unless it goes wrong. 'We can spend weeks in design meetings where most architects say, "Oh, I'm not sure about that mullion detail", says DiMambro. "But nobody who goes into the building notices the mullion detail. They go "It's too hot in here", or

LABORATORIES 5: INTERNATIONAL COLLABORATION

Physicists need very large particle accelerators in order to study the fundamental constituents of matter, but in the post-war years no single European country could afford to build one. The Conseil Européen pour le Recherche Nucléaire (CERN) established the European Laboratory for Particle Physics in Geneva in 1954, with funding from 12 European countries. Its many discoveries were crowned in 2012 by the discovery of the elusive Higgs boson.

"It's not bright enough". They notice whether there's somewhere to plug your laptop in, whether it's freezing cold, whether the toilets flush properly – boring stuff like that, but that's their experience of the building.'

At around the same time Ian Ritchie appointed another crucial consultant. David Kelly is a former research scientist and a veterinarian who now works all over the world as a laboratory consultant. 'Ian recognised that as the team had no neuroscience lab experience they needed some advice', says Kelly. 'I immediately gelled with the IRAL team. I've always drawn and painted quite seriously, and my wife's an architect. I'm not a dinosaur or a philistine. I want to work with people who want to create a nice space and a good building. I work mostly side by side with [senior project architect] Chris Russell – we just seem to understand what the other's thinking. He's very quick on the uptake. It's been a very good relationship, thoroughly enjoyable. Sometimes when you are regarded as an expert people take as gospel everything you say, and that's not very satisfactory. But Chris would say "Why are you doing it like that, why haven't you thought of doing it like this?" I very quickly realised that I would have to sharpen my way of thinking.'

Kelly brought 30 years of experience of changing trends in laboratory design, as well as detailed understanding of the stringent requirements for the health and welfare of both research workers and laboratory animals. 'There was very little change in laboratories until the 1960s,' he says. 'The traditional layout was long rows of dark mahogany benches. The Health and Safety at Work act in the 1970s

Typical neuroscience lab environment (here at the University of Edinburgh), showing high demand for electrical sockets was the main driver for change. It led to more things being done in fume cupboards or in safety cabinets. The sort of thing that as a vet student I did on an open bench, anybody doing that now would be in serious trouble.'

There was no particular moment at which the world of laboratory design realised that an acre of fixed benches was not the best idea, but a number of factors have combined to make a regular reassessment inevitable. 'By the time we got into the 1990s we were thinking more about equipment and people', says Kelly. 'That led to more and more use of dedicated equipment space: centrifuges or things that gave off a lot of heat would go into a separate equipment room. One of the biggest changes is that everything is becoming miniaturised. A lot of the analytical equipment, mass spectrometry and so on, is getting smaller. I'm allowing about a third less bench space for equipment than I would have done ten years ago.'

Meanwhile researchers now spend at least as much time at their computers, analysing data and writing papers, as they do at the laboratory bench. Health and safety considerations make it inadvisable to carry laptops backwards and forwards between the two. 'They need bench space and they need write-up space', says Kelly. 'Immediately you've got an average temporal utilisation of space of 50 per cent. How do you improve this efficiency? You can look at hot-benching and hot-desking. But when you talk to scientists, having their own space with a picture of the dog and the cat and the children is absolutely critical. I've seen hot desking done and I don't think it's a good idea. If a scientist is just a number, will he or she talk to the next number? It's important to keep this human element.'

At the same time, Kelly points out that the 'opportunity window' within which it is possible to design a building can be relatively constrained. 'You've got to consider several questions. Is it affordable? Is it compliant (with planning, animal welfare and health and safety requirements)? Is it sustainable? Is it acceptable? Is it adaptable? It's where all those things come together that drives the final form of the building. But if you can make it as adaptable as possible then it makes everything else a lot easier.' The approach he favours is to segregate within the building special functions such as animal facilities and biological containment suites that have rigid legislative constraints, which then allows the architects to have a much freer hand with everything else.

LABORATORIES 6:
THE MRC LMB
At Cambridge University in the 1950s, James Watson and Francis Crick solved the molecular structure of DNA while working in a small group led by Max Perutz. In 1962 the Medical Research Council opened the Laboratory of Molecular Biology (LMB) in Cambridge under Perutz's chairmanship. Tea and lunch breaks in the staff canteen were central to Perutz's collaborative and interdisciplinary approach, which to date has produced 13 Nobel prizewinners.

At the end of November 2009, Sarah Caddick led a team of six on a mission to California. Ian Ritchie and his colleague Loke Shee Ming, together with Dave Scott and Richard Morris from the Wellcome Trust and the project sponsor Stuart Johnson, came to visit laboratories built in the past few decades, all housing world-class neuroscientists, to see what they could learn.

Caddick, who earlier in her career had seen the kind of pitfalls that could open up if architects were too wrapped up in their own 'vision' of a building, was excited by the open-mindedness with which the SWC design began. 'To be faced with this project where the intent was to start from the inside and work out was remarkable', she says. 'Ian and his team said, "We envisage these types of challenges, having to listen to these types of people, having to visit and learn the history of different buildings - and having both to embrace and dismiss what scientists will tell you."'

'We had to understand how neuroscientists work', says Ritchie. 'There are 100,000 today, maybe there will be a quarter of a million in the next few years. Predicting how they will use space is difficult to grasp, but equally we felt we could find common denominators that were both spatial and technical.' He was not interested in traditional lab design, but wanted to start afresh. 'Among architects there appears to be a received notion of what a lab is in terms of its physicality', he says. 'A football field of benches, and, if you're lucky, an atrium in the middle. We didn't want to come with any preconceptions. Understanding how neuroscientists work, and could work, was the first move. We were very lucky: we got six months to a year to go and visit a lot of labs and talk to neuroscientists, and discover what makes them fascinated by their subject. I think John O'Keefe and his colleagues were also trying to work out who we were, and that exchange was quite rich. Particularly at the level of language – many architects speak archibabble half the time, neuroscientists speak their own language. There was a genuine enjoyment in trying to understand each other's approach to the design of this building.'

Responsibility for choosing which labs to visit fell to Caddick, whose first-hand knowledge of the neuroscience scene on both sides of the Atlantic was invaluable. First on her list was the Salk Institute for Biological Studies in La Jolla, near San Diego. A man of great aesthetic sensibility, Jonas Salk told his architect Louis Kahn to 'create a facility worthy of a visit by Picasso'. Completed in 1963, the result was a pair of tawny concrete, glass and teak pavilions that face each other across an expanse of marble courtyard, through which a shining ribbon of water flows toward the sea. Full height glass panels fill the

Circulation space at the Salk Institute,
La Jolla, California

laboratories with natural light. These labs are open plan, designed to be adaptable as needs change, and to foster interactions both within and between research groups. Individual offices for senior staff are built into towers that project into the courtyard: they have natural wood floors and views directly out over the Pacific, so that the play of light on sky and sea is a constant focus of contemplation and inspiration.

To support this openness and flexibility, Kahn was the first in the world to adopt the concept of an 'interstitial space' – an 8-foot high space above the laboratory floors that contains all the services. This allows maintenance workers to change or service technical plant such as air handling or the delivery of laboratory gases without having to close the labs themselves. The interstitial spaces also conceal rigid steel trusses that support the building so that there is no need for columns in the open lab floors.

'Going to the Salk 46 years after it was built and seeing how it had adapted to the challenges of changing techniques was a revelation', says Ritchie. 'The interstitial floor is clearly a huge bonus. The separation between laboratory and write up areas, the climate, the scale of space, the idea that you could go into your beehive monastery cell: I thought those were really nice. It is one of the most beautiful pieces of 20th century architecture. It will last. Louis Kahn's concept is beautifully thought through.'

Richard Morris, as the only working neuroscientist among the visitors, was also suitably impressed. 'We realised that the thin walls on the main lab floor were perfectly moveable: we had not previously understood that there was that level of flexibility', he says. 'Of course, on reflection, it's obvious that you need that. Microscopists need lots of small, dark rooms, biochemists don't.' He did, however, have one significant criticism. 'One thing that was for me very important was that the animal house is away from the main building and also underground. All the labs that do any kind of *in vivo* work, whether it's behaviour, electrophysiology or microscopy, will send people there and set up experimental rigs there, underground, away from daylight. So every lab is split, and nobody liked that.'

At the time of the Glimmer team's visit, the Institute housed between 1200 and 1500 scientists, students and technicians, investigating fundamental questions in neuroscience, genetics, development, viral infection, cancer, diabetes and plant biology. Ritchie made detailed notes of the dimensions, materials and furnishings of the laboratories and office spaces. But he was equally interested in how the Salk's scientists responded to the space they occupied. Professor Fred 'Rusty' Gage works

Typically cluttered lab interior, here at the Salk Institute

on the growth of new brain cells in adults. As well as his lab, he showed the visitors his study, a retreat from the bustle at the bench.

In Fred Gage, Ritchie immediately found someone who shared his excitement about the intersection between neuroscience and architecture. 'I asked the question, "As an architect, do I know how to design a staircase?"', says Ritchie. 'I know the geometry of a staircase, I know the rules about staircases, but I don't know what the design of a staircase is. Let's assume there's a café up the staircase, and one down the staircase. Can I design a staircase that will make people go down in the evening and up in the morning? I thought that if we got together, we might find an idea. So we talked about light levels, the width of staircase, whether you taper them.' Ritchie also spent a lot of time talking to John Reynolds, the professor of systems neurobiology, who works on the limits of attention, and who also has a particular empathy towards understanding architecture. 'I think those two people particularly impressed me when we were there, as very sane neuroscientists trying to find real answers' says Ritchie.

LABORATORIES 7: THE SALK INSTITUTE
Jonas Salk's vaccine against polio almost eradicated the disease worldwide. Given land by the city of San Diego and financial support from the National Foundation for Infantile Paralysis, he commissioned the modernist architect Louis Kahn to design an institute to study the biological basis of human and plant diseases. Opened in 1963, the Institute's priorities were adaptability to cope with future change, and the promotion of collaboration between disciplines.

The open courtyard and outside corridors at the Salk provide spaces for interaction for all the Institute's staff and students, while slate panels strategically placed near staircases served as instant blackboards to support the kind of spontaneous interactions the building was designed to encourage. Nowadays the Salk scientists scribble on the insides of the glass façades with whiteboard markers. Far from being shocked at this apparent lack of respect for the architect's vision, Ritchie immediately saw that allowing scientists to write on as many vertical surfaces as possible would make them feel at home. Later that day, at the new Center for Neural Circuits and Behavior at the University of San Diego, Ritchie noted that 'Labs have access to external terraces where researchers can have barbecues, play table-tennis, or even take naps on a hammock', emphasising the scope for playfulness within the work environment.

Ritchie came away with his mind buzzing with impressions of the kind of people he would be designing for. 'Researchers were very articulate and highly intelligent', he says. 'But a lot of the spaces in the labs were filled up with stuff. One person in San Diego even had his dog in there. They were quirky. That quirkiness doesn't lend itself to generic solutions. In neuroscience the groups are quite small. The image one has of a laboratory is of rows and rows of people at benches, like

clerks in an insurance company, but it was nothing like that. That was a nice thing that we discovered: that they are all in a playground, or a sandpit.'

If the scientists are in a playground, then they have demanding toys. Their aim is to make the connection between the activity that goes on in the brain, and the behaviour of living organisms. While traditional anatomy – slices of brain exposed under a microscope - can reveal how one cell is connected to another, only an active, living brain can show how those cells work together as circuits. Changes take place at every level, from the genes that provide the instruction books for the cells, to the behaving organism: a range of scales that encompasses seven orders of magnitude. Such complexity demands an inventive array of technology, using biochemistry to label molecular interactions, monitoring electrical signals in small brain areas, the imaging of whole brains, and recording the activity of animals as they complete a variety of behavioural tasks.

'I realised that if you can put 1000 sockets in a room they will love you' says Ritchie. 'Everything is electrically powered. No lab looks the same – they are all completely different. Some were just closets off the side of a corridor. And I had a sense that most of the kit was built by the neuroscientists themselves. It was nothing to do with management teams coming in from a university.'

The team flew north the next day and visited three labs at the massive new development for the University of California at San Francisco on Mission Bay. In Genentech Hall, designed to accommodate 100 research groups, they found the labs organised into clusters or 'neighbourhoods' arranged around a shared kitchen/

Space for relaxing at the University of San Diego

coffee room and offices. The coffee room is clearly an important hub for informal meetings and conversations, and researchers often gather there to eat their packed lunches together. Few of the institutions the team visited in California provide a cafeteria on site: people either bring their own lunch, or get in their cars and drive to a restaurant.

Other buildings on the Mission Bay site, clearly built to severe cost and time constraints, contrasted unfavourably with the Salk's attention to detail and future-proofing. 'In one of the buildings we looked at, the suspended ceilings were half done and the cabling was just thrown in – it was spaghetti chaos', says Ritchie. Even in a building that had been more sympathetically designed, labs had been fitted with benches that already had to be modified because they were not needed for the science in those labs. 'We realised that you can make a mistake' says Richard Morris. 'New is not always better – new can be worse.'

The California visit was the beginning of a programme of laboratory visits that continued well into the following year. Included were Harvard, MIT and Columbia Universities in the US; the Champalimaud Centre for the Unknown in Lisbon, Portugal; and in the UK, Edinburgh University, Oxford University and the Sanger Institute near Cambridge, plus a number of UCL laboratories.

Many of these had been built with collaboration and adaptability in mind: some had achieved it more successfully than others, as Chris Russell remembers. At Harvard he visited the laboratory of Josh Sanes and Jeff Lichtman, creators of

Ian Ritchie (right) visits a lab at the University of San Diego

the 'Brainbow' technology that labels different brain activities with a wide range of fluorescent colours. 'They were in a fairly new building', says Russell. 'Previously their offices were accessed through the lab: in the new facility all the offices are on the opposite side of the building from the lab. The PIs' [principal investigators'] office has an antespace for a secretary or assistant, then a corridor, a support space, and finally then there's the lab. The PIs hardly ever see the people in their lab any more. Before, they would walk through the lab to get to their office and people would stop them to have conversations; now they were behind a gatekeeper across the corridor. They suddenly became psychologically unapproachable when they moved to the new lab.'

At the other extreme, conversations took place that appeared to suggest that for some, architecture was almost irrelevant. Ian Ritchie went to see Richard Axel at Columbia, who had inspired David Sainsbury to get more involved in the field years before, and was now on the Governing Council of the projected laboratory at UCL. 'I was struck by the fact that he had Goethe and a book on opera next to each other on his shelves', says Ritchie. 'Both he and Eric Kandel had a lot of space. What was fascinating about Richard is that you could ask big questions and he gave very straightforward answers. He said "You've got to realise that this work is incredibly boring – 95 per cent of the time it's just logical or repetitive. Fortunately

Writeable surfaces and write-up area at Harvard University

I don't have to do that, the post-docs do it." The lab space where they were working was not important to him any more.' What was important to Axel was conversations – it is the quality of these interactions that matters.

'He gave me a much more open-ended idea of what a lab could be', says Ritchie. 'This idea that you have teams of 9–12 working with PIs – it made me think "Is that the right approach?" Architectural ownership is irrelevant. It's their space, they can do what they like.' Another Columbia scientist, Tom Jessell, was refitting a laboratory for a colleague newly arrived from elsewhere. 'He was gutting a lab in an old building completely. He wanted empty corridors going round an island in the middle. That was where I realised that many neuroscience labs don't like a lot of daylight. They need darkrooms where they can use optical instruments. They would put up a black cotton curtain, on rails, so that you'd have very different experiments next to each other.'

The architects carefully documented each visit, photographically and through detailed notes. By the end of the process, they were beginning to formulate a strategy for how to design the building, and had an even clearer idea of how not to do it.

Super-resolution fluorescence microscopy at Harvard

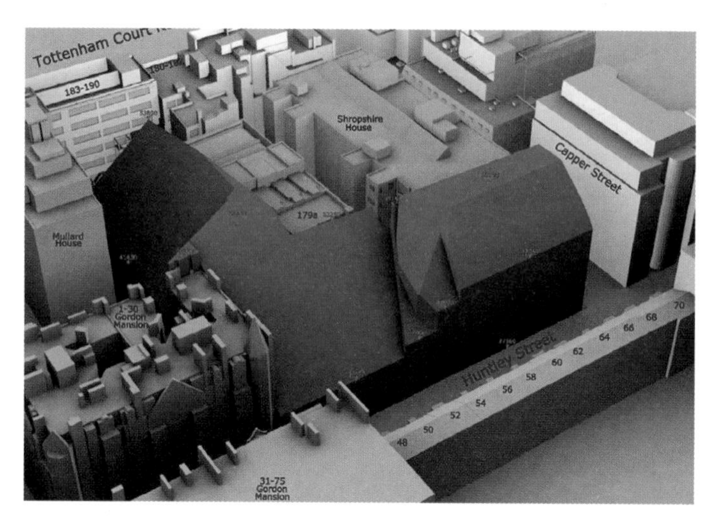

Back on the ground in London, Ian Ritchie's concerns about the Huntley Street site had proved to be well founded. He and his colleagues at IRAL made a detailed study and by January 2010 delivered a 'Stage A/B' (appraisal and design brief) report that convinced Stuart Johnson they would have to find another site. 'Ian was very shrewd on the site constraints', he says. The existing building in Huntley Street was only one-and-a-half storeys high. There was residential accommodation nearby, and building to a height that would meet the space requirements of the brief would have an unacceptable impact on the daylight and sunlight available to the neighbours. 'They tried weird-shaped buildings to get round the problem, but it was completely unfeasible. We would not have got planning permission,' says Johnson.

With the support of the funders, he took the decision to suspend the project until the Provost of UCL had come up with another site. He had just two months. After considering all the options Malcolm Grant offered the site of the Windeyer Building in Howland Street, originally the home of the Middlesex Hospital Medical School and at that time still occupied by UCL's department of infection and immunity. 'Another complication', says Grant, 'was that there was also activity in the building from University College London Hospital. There were really quite tricky issues we had to resolve with them. It was by and large amicable, but it took a while to come up with plans on a cost-sharing basis that both parties thought equitable. It was a bit touch and go.'

Left Weird-shaped solution to the massing problem on the Huntley Street site.
Right Alternative site: the Windeyer Building on Howland Street

Playing a key part in implementing this decision was Dave Smith, deputy director of estates at UCL with responsibility for space allocation. 'We had not originally planned to demolish the Windeyer', he says. 'It was always a building that was at risk, but it's such a big building – 13,500 m^2 full of science – and we had nowhere to put them on the face of it. It is a site that is highly valuable in the area.' Once UCL offered the Windeyer site the SWC team eagerly accepted it, but with a limit on any further delay. 'Then it became one helluva challenge to meet the timescale', says Smith. 'The institution saw this project as worth knocking a building down, and moving people all over the estate. Some have gone up to the Royal Free, many have gone into the Cruciform Building, and others are scattered around elsewhere. But Malcolm Grant was very clear we had to do this.'

Throughout this period IRAL led the design team on a series of building studies, including one that was not tied to a particular site. 'The theoretical building was a generic, "If you put a neuroscience lab on an unrestricted site, where would all the pieces go?"', says Chris Russell. 'We had a bit of freedom to think about how the different parts would fit together. We were then given three possible sites, and we did feasibility studies for all three. The Howland Street site was the one we were all most comfortable that we could fit the building on in an organised way.'

The design team also capitalised on the enforced delay by undertaking more detailed research. It began on 27 January 2010 with a neurosciences interaction day, organised and introduced by Sarah Caddick, and billed as 'Educating the Architect's Team'. The idea was to brief the team about the spectrum of work that would take place in the new institute. In her introduction, Caddick said that she hoped the day would inspire the design team to look beyond 'space metrics' and deliver a building that is 'creative, useful, adaptable and wonderful'.

Six senior UCL neuroscientists, including John O'Keefe and Peter Dayan, as well as Richard Morris from Wellcome, then gave unusual presentations about their work. Their brief was to address their branch of neuroscience and why it was important; the infrastructure needs in a new laboratory; and likely changes over the following decade. Some neuroscientists work at the level of individual molecules, on a scale of nanometres. Others deal with whole organisms or neural pathways measured in metres. The building would have to accommodate them all. The talks were

BRAIN SCALE 1: NANOMETRE TO MICROMETRE ($1X10^{-9}$ – $1X10^{-6}$ m)
The synaptic cleft between one neuron and another is about 20 nm wide, or one 5000th of the width of a human hair. Protein complexes embedded in the cell membrane either side of the cleft help to stabilise the synapse or act as neurotransmitter receptors. Working at this scale requires ultra-high resolution imaging, biochemistry and genetic manipulation.

organised to progress from the smallest scale to the largest, covering a similarly wide range of timescales.

Whereas a scientific presentation might normally follow the conventional route of aims, methods, results and conclusions, these researchers focused on how they worked, the space they occupied, the social interactions within and between groups, the requirements for specialised scientific instruments, and the housing of laboratory animals from zebrafish to mice and rats. A picture emerged of a society dedicated to a similar end – understanding how the brain works – but organised into autonomous groups with a great deal of flexibility about how they arranged themselves in space.

To Chris Russell, the event was hugely informative. 'It was fantastic', he says. 'No one tried to say what neuroscience is. We had a series of completely fascinating, very short presentations from the neuroscientists, and not one of them was the same. They spoke in order of scale, so Yukiko [Goda] started at the nanometre, and she described the process that allowed her team to study the way two or three neurons communicate at a microscopic level. We continued to learn more with each presentation.'

'It was a very inspiring day', says Gordon Talbot. 'For us the key thing is what direction the science might go in the future. We endlessly harassed Sarah and John about this, and they said, "We really don't know – that's why we've employed you."'

Following the meeting the architects asked for the opportunity to spend a full day in each of the six UCL laboratories, so that they could really understand a 'day in the life' of each. 'I was meant to be a fly on the wall, just observing,' says Russell, 'but it was very difficult because the scientists wanted to explain to us what they were doing, which led to me asking more questions. They ended up being more extended tours. We also looked at the support spaces, such as shared, 'bookable' microscope rooms, that we hadn't had a chance to see on the short visits to labs. We went into the basement plantrooms, went to see where the refuse is stored and the glass washed, we spoke to the staff in the stores - we had a really broad look at everything, recorded it, asked what worked and what didn't, what they wished they had and what they'd like to do more easily. With each of those visits, we increased our understanding of neuroscience.'

On these 'day in the life' visits the architects got the chance to talk to the real 'workers' – the graduate students and postdocs – and not just the group

BRAIN SCALE 2: MICROMETRE TO MILLIMETRE ($1X10^{-6}$ – $1X10^{-3}$ m)
There are many different populations of neurons in the brain, measuring from 5–100 μm across the cell body. Their structure and connections are visible under the light microscope. Working with developing organisms, such as the zebrafish embryo, scientists can track the processes that underlie the growth and patterning of connections between cells in the nervous system.

leaders or principal investigators. 'I never brought lunch, so I would be invited by the PI, a postdoc or a PhD student to join their usual lunch routine, where I answered as many questions about architects and architecture as I asked about scientists and neuroscience', says Russell. 'We learned even more about how a university research hierarchy works. Quite a lot of the younger people come in later, and work till late. If you have flexible hours and can do it, the younger you are, the later you wake up, seems to be the rule. We gained insights about the PI's relationship with the postdocs and PhD students. Michael Häusser had an office in the lab and an office in the admin area of the building that he never used: eventually he moved five of his PhD students into it.'

'There was one visit, to Matteo Carandini's lab at the Institute of Ophthalmology, where they had quite complex behavioural equipment set up', says Talbot. 'Matteo took us aside and said "What we need in labs are these things" – and showed us steel channels supporting cable baskets from the ceiling that are stock-in-trade for us from transport and infrastructure. They're called Halfen or Unistrut channels, and they allowed him to construct platforms for equipment suspended above the experimental area, or rig, and to organise cables between the rig and computers. We immediately adopted that idea. It's an example of something that grew from a single comment and

BRAIN SCALE 3: MILLIMETRE TO METRE ($1X10^{-3}$ m – 1 m)
The axons of motor neurons that communicate between the brain and spinal cord, or spinal cord and muscles, can be over 1 m in length. Sensory signals registering as touch, pain, heat, pressure and so on have to travel similar distances from the body surface to the brain. On this scale scientists are considering the organism as an integrated whole, with many factors contributing to the behaviour that results from any individual stimulus.

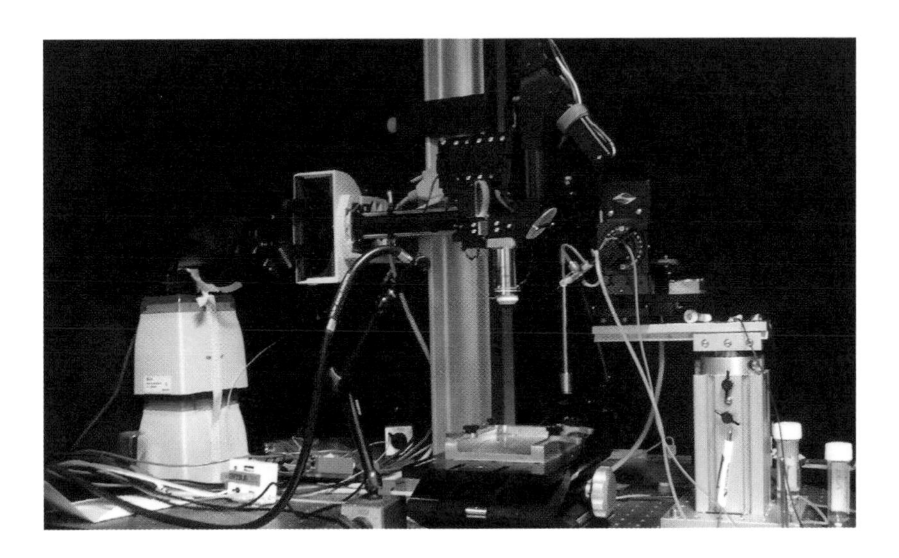

Two-photon microscope in Michael Häusser's lab at UCL

coalesced a discussion. Otherwise we might not have thought of including them as part of the building's physical infrastructure that could also be used by the scientists.'

One of those who hosted a 'day in the life' visit was Marg Glover, who was then interim centre manager for the SWC and would go on to play a key role in advising on practical, day-to-day matters as the design developed. 'We had a "day in the life" to identify the sorts of things that are going to be required for administration in the building', she says. 'It was very valuable to have those discussions. They are looking at the aesthetics. I'm at the other extreme – it just has to be practical as far as I'm concerned. I'm just delighted when they take one of my ideas on board.'

'The breadth of access we were given was very important', continues Russell. 'It stopped us having just one scientific voice that doesn't necessarily represent everybody. Everything from how they work to what kind of space they need, whether they prefer open plan write-up, or cellular office write-ups: everybody had a different opinion. It was also extremely useful to understand how the non-scientific support staff work, and to hear their opinions.'

BRAIN SCALE 4:
OVER A METRE
Ultimately the study of behaviour has to take into account the animal's environment, including the space in which it moves. Undertaking studies at this scale is particularly challenging because of the need to maintain strict control over the conditions under which the experiment is performed.

One way of providing space for interaction: the Cruciform Building, UCL

The engineers from Arup also took the opportunity to visit labs. 'To see physically what they've built is not particularly helpful from a structural engineering point of view', says Christian Allison. 'What is interesting is to talk to people about how they use the building. As a team we are trying to make a building that's not only working from a performance point of view, but from a human point of view. We would learn something if we were going to look at a new system or new product. But other than that it's more about the users.' Jennifer DiMambro agrees. 'Quite often it's the operability', she says. 'We were shown round by the building manager, and we'd sometimes learn what not to do. This kind of visit is most useful for the architects, especially if they haven't done a lab before. It's also good for the end users to see other environments, because often they've been in the same space for a long time. They don't have the experience of different environments to be able to comment.'

Space for formal or informal interaction between members of the laboratory had emerged as an important consideration on all the lab tours. The older labs at UCL had varying degrees of interaction space. 'In the Cruciform Building on the way to the cafeteria they actually had a sign up that said "Interactive Wing"', says Russell. 'Which is basically two meeting rooms and a cafeteria. But when we went

An alternative: break-out space off circulation corridor at the Informatics Building, University of Edinburgh

to the Informatics Building in Edinburgh with Richard Morris, there was quite a lot of engineering of spaces for people just to sit and meet and gather.' Having a white board or other writeable surface to hand in these spaces was also seen as highly desirable. 'When two PIs meet in the lift or the corridor, they sometimes need to start writing formulas or diagrams to demonstrate what they're discussing', says Russell. 'As architects, we completely understand that you can't just wave your hands and describe something that might be complex spatially, such as the way things connect together.'

The process of exchange continued well into the design phase of the building. A year after the neurosciences interaction day – a period in which the UCL neuroscience community had continued to build its own sense of unity – Ian Ritchie and John O'Keefe presented initial ideas for the Sainsbury Wellcome Centre to them and to senior figures including the Provost. Ritchie's talk expressed his appreciation and understanding to date of how neuroscientists' spatial, technical and social demands had been gleaned from the many visits. After the presentations there was a general discussion between the scientists and the design team. Some of the scientists' requirements seemed contradictory. Everyone agreed that social interaction was important to keep ideas flowing, but some also yearned for private

Outdoor theatre overlooking the sea at
Champalimaud Centre for the Unknown, Lisbon

spaces, like the cells of monks, where they could think without interruption. The experimental neuroscientists were determined that their subjects should be housed close to the laboratory, and not in some distant Biological Services Unit (BSU). But Home Office regulations, designed to safeguard the health and welfare of animals used in research, can make this difficult to achieve. All agreed that new technologies that might become available in the future could change the requirements for space and services.

'One scientist suddenly said, "We want opening windows"', says Ian Ritchie. 'Somebody else said "You must be joking, you open the window, spill the mercury, it kills the people outside...", and so there started quite a lively debate about what a lab is.' His own understanding of what makes neuroscientists tick continued to deepen at a 3-day symposium in October 2011 at the Royal Academy of Medical Science, organised by Gatsby to bring together neuroscientists it had funded from Columbia University, London, and the Weizmann Institute in Jerusalem. One presentation showed how the brain responds differently to abstract art, measured using a magnetic resonance imaging (MRI) scanner, compared with figurative painting. 'If you showed them an image of a tree or a face, it activates the same part of the brain in all subjects', Ritchie remembers. 'Put in an abstract painting, and it's all over the place.'

Richard Axel also gave a paper about what we know about where we're going. 'He used some delightful art illustrations, which I was very struck by', says Ritchie. 'His knowledge of art history was a surprise. When I saw John O'Keefe give a paper at the Royal Society, he again used art illustrations. I was beginning to feel that the idea of giving them a sandpit was the right one. Underneath we are looking at people who are looking at the brain in a broader way than normal scientists. I was beginning to really enjoy these people.'

3

INTO THE WORKSHOP

SCIENCE AND ARCHITECTURE IN DIALOGUE

On Friday 4 June 2010, Project Glimmer's first 'Workshop with Scientific Advisers' took place in one of the meeting rooms at the Wellcome Trust's new headquarters at 215 Euston Road. The ten-storey, open-plan, glass-roofed Gibbs Building, designed by Hopkins Architects, was opened by Her Majesty the Queen in December 2004. Seated round the boardroom table were representatives of all the project's many constituencies: the scientists, the architects, the engineers, the funders, the project managers, the UCL and Wellcome estates managers, and a range of specialist consultants.

The programme of workshops was agreed between the architects and project sponsor Stuart Johnson soon after the neurosciences interaction day showed how fruitful such discussions could be. The idea was to take all the interested parties along as the building's design developed, with a two-way dialogue ensuring that there were no nasty surprises on either side. On their visits to other labs, the design team had encountered several examples of neuroscientists who had been presented with new labs without ever being asked for their input. Often the things they really needed had been left out, and things they didn't need had been provided at great expense. The workshops provided a means for the scientists to offer critical input at every stage of the design, and to understand the issues that the designers would be facing as they created the building.

'I like process, and I like the hierarchy of meetings', says Stuart Johnson, who organised the programme of workshops, drew up an agenda for each in consultation with the design team, and took the chair at each meeting. 'I don't like filling people's diaries. I like meetings to have a point, and to vary over the life of the project. So we planned these meetings on a regular basis – once a month in the early stages, twice a year later on – that bring the funders and the science advisers and the design team together to have presentations and if appropriate, sign off on particular issues.'

An unusual feature of the SWC project was that apart from John O'Keefe and Peter Dayan, the lab's users were a completely unknown quantity. 'We haven't had the individual PIs to go to', says David Kelly. 'Sometimes that can be a mixed blessing, because (with a strong end user) you end up with something very prescriptive designed round an individual. It's quite a challenge designing facilities without an end user client, but in terms of adaptability you stand a chance of getting a better building. The workshop process has been very productive.'

While Stuart Johnson presented the workshop programme as a tool for decision-making, it had another function that may well have made everything else

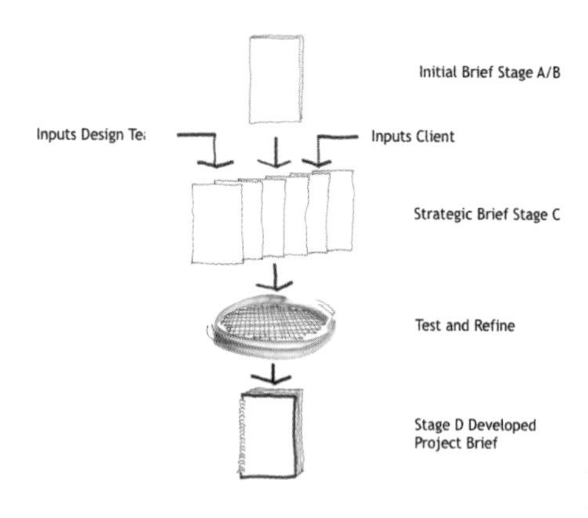

RIBA OUTLINE PLAN OF WORK
The Work Stages specified by the Royal Institute of British Architects (RIBA) run from Appraisal (A) and Design Brief (B) all the way to Post Practical Completion (L). Critical design stages include Design Development (D), at which point the design is submitted for detailed planning permission, and pre-construction stages F, G and H, which lead up to the appointment of trade and specialist contractors. In 2013, the RIBA work stages were amended to 1 Preparation, 2 Concept Design, 3 Developed Design, 4 Technical Design, 5 Specialist Design, 6 Construction (offsite and onsite), 7 Use & Aftercare.

possible. No one was more aware than Susie Sainsbury (who attended all the meetings) that the large team collaborating on the project had different tribal affiliations and even different languages. 'There were two different vocabularies in use', she says. 'One that the neuroscientists used, which was not fully understood by architects, and vice-versa. I realised we could end up with a terrible muddle, because they might as well have been talking Lithuanian and we were talking ancient Celtic. The workshops changed that; now the languages are shared. It's been quite exciting just to watch them using each other's vocabulary. You hear a young architect chatting about neurons when he wouldn't have known what a neuron was if it had come up and hit him two or three years ago.'

The design team gradually became aware how little the scientists understood about the design process, and that they needed to explain things that seemed absolutely obvious. 'For the neuroscientists, it's about understanding when they need to flag up a concern', says Susie Sainsbury. 'A majority of the scientists can't read pictures the way the architects can. Now they understand a lot more about why you have to take a decision today about something that won't exist for three or four years.'

After the interaction day and the lab visits, which were all about the architects finding out about neuroscience, the workshops turned the tables. 'The early days were just learning seminars in which we had to get up to speed on what the constraints were on building a building like this', says John O'Keefe, who then held the title of Interim Director. 'I didn't have any previous experience of working with architects, though I'm a great fan of architecture. I grew up in New York and remember when I first saw the Seagram building, the only Mies van de Rohe building in New York. So I knew what I was getting into, and I've been very pleasantly surprised. Ian and I get along very well – I appreciate his artistic side. He views himself as an artist although he's a very good engineer and a very good architect.'

Gordon Talbot, the project director from IRAL, led the workshop presentations on the design as the project progressed. Talbot is a long-standing member of the practice who has previously worked on technically complex projects for London

Right Thinking about lab spaces

Underground and Network Rail. His role has been to develop Ian Ritchie's vision for the project into concrete reality. 'I think about how we are going to answer the questions that raises, the means by which we do it, what we need to find to tell us how to design the building', he says. 'Then I map how all that comes together, not only in our internal team, but everything that the other disciplines are going to do: structures, MEP, lab consultant, planning and building control.'

This kind of advance planning is essential to the success of a project. 'When you are within a week of having to produce a report, it's way way too late to decide that we needed to go and talk to these people, agree this with the client and so on', says Talbot. 'So tracking what we're doing, how we're doing it and when we're doing it is a really important part of my role. There's a big chunk of design, a big chunk of making sure that we're doing what Ian wants and what Ian is telling the client they're going to get, and then there's making it all happen.' Talbot prepared and circulated reports on all the workshops, so that every comment and decision was documented.

At first he and his colleagues had no idea how the workshop process would work out. 'The first one of these meetings, we weren't really sure who was presenting,' he says. 'And on the way there in the taxi Ian said, "I'm not going to have anything to do with this, I'm going to sit with the client group." So Chris, Loke and I started presenting our material, and then Ian started heckling. Which is very funny, because it puts us in the position where we have to explain more clearly what we were doing. And gradually the scientists and funders and client group became more confident about asking questions, so it was a really successful move.' Subsequent workshops followed the same formula, with members of the design team presenting their own areas of responsibility, while Ritchie sat with the client group.

If the scientists thought they were going to see an arty picture of the new building at the first session they were mistaken. In the eight months since they had been appointed, and informed by their visits to labs around the world, the design team had been working on the principles of a theoretical building that

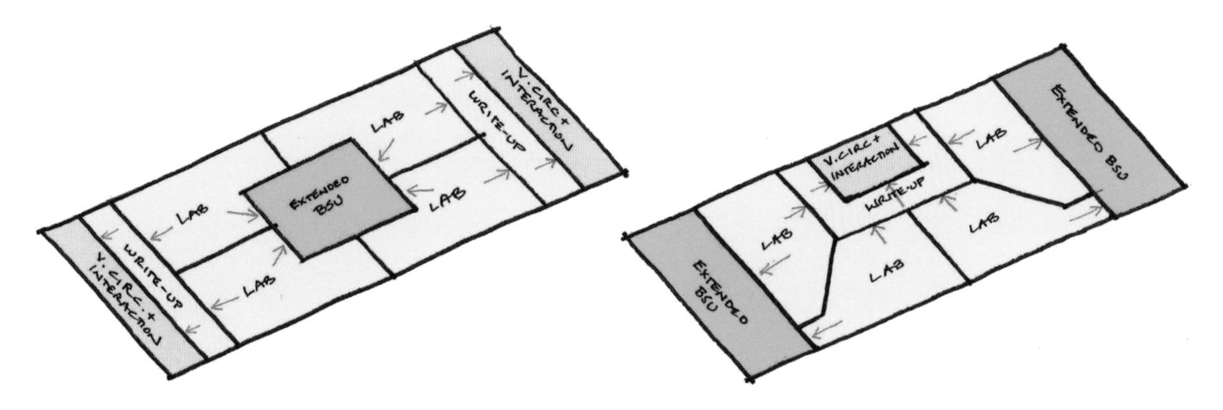

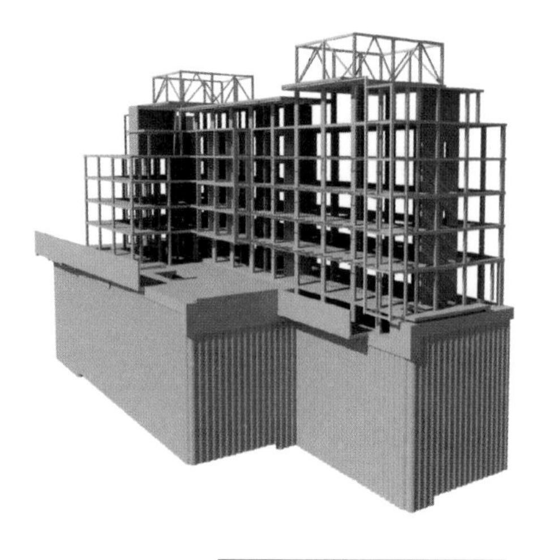

CONCRETE FRAME STRUCTURES
Most modern office buildings are based on a steel or concrete frame. The overall dimensions of the frame are determined by the massing possible on the site and the square footage required by the design brief. The spacing of the grid of columns and beams that make up the frame is a crucial feature of the design, as it will determine the volumes of the spaces inside the building as well as ensuring that the building is structurally sound. The frame is supported on foundations sunk into the ground.

would meet the needs of the scientists while conforming to the requirements of budget, local environment, energy efficiency and so on. This theoretical building provided the basis for a report on the Howland Street site that the architects entitled 'Stage A/B Revisited' – the original Stage A/B report having concerned the rejected Huntley Street site.

In this work IRAL was closely supported by the Arup engineers. 'In the early stages we were effectively providing an advisory service to Ian's team,' says Christian Allison, 'to help them develop serviced floor layouts that are going to work. Floor-to-ceiling heights for example: we spent ages talking about how to get the lab services to work, what slab thicknesses we need structurally to provide the vibration performance and so on.' Jennifer DiMambro adds 'We understand what they are trying to achieve, but we have to make it work so that when the scientists come in they can access the services, or the vibration is adequately controlled. A compromise sounds like a negative thing, but it's a balance.'

The overall footprint and height of the building were largely defined by the size of the site and the London strategic view corridors dictated by local planning considerations. But at the beginning the architects kept an open mind about the internal layout, and made no attempt to delineate the face that the building would present to the world. Instead they addressed the issues that had arisen from the lab visits, and showed the workshop participants a variety of possible solutions.

The scientists quickly learned that every option had consequences that might affect the structure, the services, and ultimately the cost of the building, as well as the laboratory environment. The funding provided for the building by Gatsby and Wellcome was generous, but not unlimited. Cost was always a factor, but the demands of the science were paramount. For example, the architects and engineers paid much more attention to vibration and electromagnetic compatibility than they would have done in a conventional office or laboratory building. Any electrical interference can ruin an experiment involving low-voltage signals or delicate electron microscopes. 'Once we got down to levels of detail about things that had gone wrong in buildings, one key requirement was getting

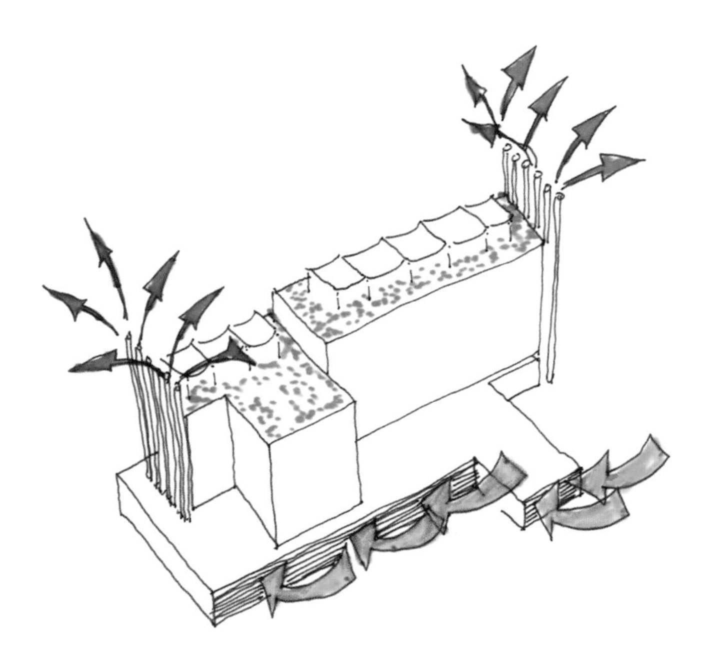

rid of the electromagnetic noise', says Chris Russell. 'In one instance we heard of high-voltage cables passing through the electron microscope room. Even in John O'Keefe's lab, the air conditioning fans for the building were on the outside wall of their lab, which didn't help.'

Other discussions centred on features that would be nice to have – and therefore enhance the working environment – rather than critical for the science. John O'Keefe particularly remembers a confrontation with Jennifer DiMambro over the roof. An early sketch of the theoretical building had shown the roof covered in grass, with an elegant pavilion in the centre: Ritchie had wanted to depart from the laboratory design convention that put plant to service the building on the roof. 'In one of the early meetings Jennifer said, "I want the blowers [exhaust fans for the ventilation system] on the roof,"' says O'Keefe. 'And we'd been thinking this roof is just the right height to have a garden with nice views. At the next meeting we were talking about how we would manage the air control, and I said, "Oh, and I want the windows to open." She said "What???" Then she thought for a second and said "OK - windows that open, blowers on the roof." So we have windows that open, and blowers up on the roof. But IRAL put them up on tables, so we still have a nice landscaped roof space, and that was a good compromise. And that's representative of a lot of things that have gone on.'

MASSING
Architects use the term 'massing' to refer to the overall size and shape of a building: its height, breadth and depth, and whether its shape is a simple rectangular box or something more complex. It is constrained by restrictions imposed by planning authorities on maximum height, distance from other buildings and so on. The architect will also take account of its proportions and its relationship to the surrounding streetscape, while providing the square footage prescribed in the client's brief.

Early architects' sketch illustrating flow of air (and desirable roof garden)

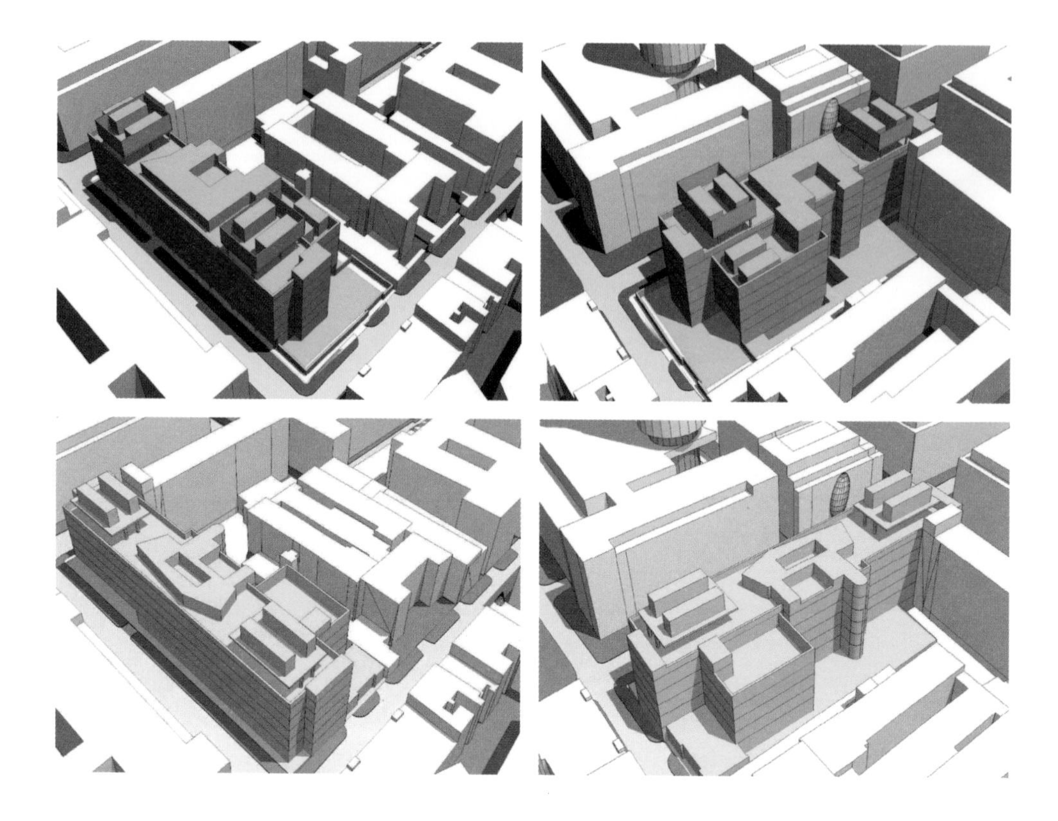

Between workshops the architects, engineers and consultants went back to the drawing board and tested the effects of different design decisions through a numbered series of building studies. Some were little more than a sketch, others worked up in detail with plans, elevations, models and analysis of the engineering aspects. Eventually there were more than two dozen. 'Not all the building studies lived long enough to be a document', says Chris Russell. 'A lot of them were just space planning exercises that were abandoned. In some instances Loke and I were each doing two building studies at the same time. The ones we thought worked and started to answer some of the issues of relationships of space and people would be discussed with Arup. Typically we had at least one sketch with coloured elevations and plans that showed how the building was organised in each of those studies.'

Christian Allison of Arup advised on the structural implications of each change. 'It starts with the massing – the overall size and shape of the building',

Two of the two dozen or so building studies created on the way to the SWC

he says. 'For each model we will have done quite a lot of work. You have to put enough thought into each one to say what the pros and cons are, and whether it's feasible. Ultimately, once you've selected what the arrangement is going to be, you have to be sufficiently confident from all engineering aspects that it will work three years down the line. It's one of the most important stages of the design process. If you get it wrong at that point, then the consequences further down the line are fairly extreme.' His colleague Jennifer DiMambro adds, 'For that reason, you tend to have quite senior people involved, who are better informed to make that judgement.' On this project the process was an extensive one – from building study 1 to building study 23c, which was eventually adopted, took a whole year.

COLUMNS, BEAMS AND SLABS

A concrete frame building is composed of three elements: columns, beams and slabs. Columns provide vertical support, beams horizontal support, and slabs support the floors. The thickness of each component depends on the demands on the building: the 'dead load', or the intrinsic weight of the building, and the 'live load', or the weight of the occupants, furniture and equipment.

The IRAL architects explained to the scientists that they would follow the outline plan of work defined by the Royal Institute of British Architects in 2007, which helps architects to organise the process of managing and designing building projects. With significant milestones agreed in advance, there was no danger that the workshop process would extend indefinitely in a tangled skein of indecision or disagreement. The first target was Stage C, or concept design, which defines the size and shape of the building and includes outline proposals for structural and building services systems. On the Glimmer project the development of Stage C occupied the months of June, July and August 2010, with a workshop taking place in each month. Stage D, which developed the concept into a scheme design ready for submission to the planning authorities, was scheduled for September 2010 until January 2011, with workshops in alternate months. While there was still some scope for modification after Stage C, at Stage D the scheme would be 'frozen' ready for submission to the planning authority and the subsequent development into a set of work packages to be handed to the contractor for construction.

In parallel with the workshops there was a programme of smaller monthly meetings for the design team, the project team and the project management group. Seconded to all three of these groups, as well as the workshops, was the 'representative user' Arifa Naeem. Now a senior research technician at UCL's Institute of Ophthalmology, she then worked as laboratory manager for the UCL neuroscientist Michael Haüsser, and her detailed familiarity with the local research community put her in a good position to find key contacts and compile information for the discussions. 'My role was to bring forward any issues from the user point

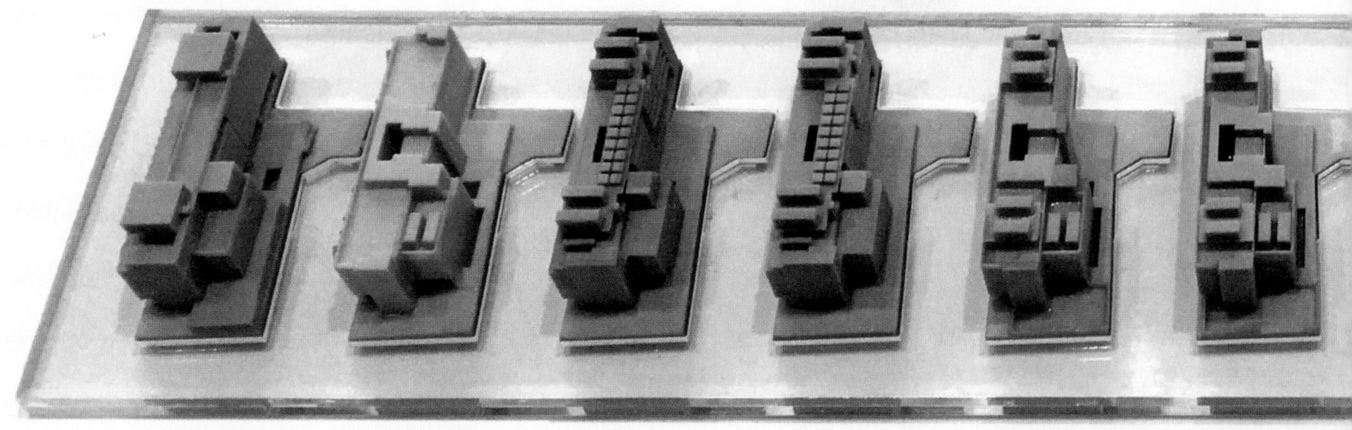

Building studies: models show the evolution
of IRAL's ideas

of view', she says, and she had considerable input into questions such as the range of equipment that might be needed, and the layout of the labs and the BSU.

Dave Smith from UCL Estates also found the process instructive. 'The workshops were very successful', he says. 'You get to meet the team in a wider sense – scientists, funders and so on – it's been quite a journey. I'm not saying it's something we've adopted, but on the more complicated jobs we do similar things, though not in such a formal way. One of the biggest issues is communication, and that solves the problem to a greater or lesser extent. Anyone can ask key players awkward questions. I think Ian and his team are very good – I would describe them almost as academics in their approach. They ask questions that I don't even know where they come from. But hopefully that will deliver a perfect building.'

Dave Scott from the Wellcome Trust attended the workshops and the project management group. 'It was a very pleasant surprise to see how Ian dissects a brief', he says. 'He literally takes everything apart, right down to the last screw and bolt. The other thing was how he successfully pulled all the diverse parties together. It's very difficult to have scientists, construction people, architects, and funders all together to agree everything at the same time. Ian very carefully structured these workshops over half a day, so the whole team could see the thought process on a range of subjects. They were glorious stepping-stones from the feasibility studies, to detailed design to construction design and implementation. If I could summarise Ian's role, it's being able to take all those diverse people along this journey.'

Supporting the design team every step of the way on this journey was Susie Sainsbury. Looking back, she can list the contentious points on which she exercised her powers of persuasion, almost all of which related to the building as a social space: 'Protecting what seemed to be a radical idea of having fewer, smaller private offices and more open space. Ensuring that ways of moving about the building push people together. Persuading Peter Dayan to have his department right in the middle: I think he still hasn't realised quite how dangerous it is. When they put their noses out of their burrows they are going to see other people! The staircase is positioned to encourage a collision of ideas. When you look at what someone else has written on their whiteboard, that's when you have ideas.'

Susie Sainsbury was also anxious that the team members should feel comfortable with each other socially, removed from the formality of meetings. 'We did a certain amount of eccentric team-building', she remembers. The rebuilding of the Royal Shakespeare Theatre was still occupying a lot of her time. So that the

Bonding exercise: design team, clients and scientific advisers go backstage at the Royal Shakespeare Theatre

Architect Gordon Talbot's sketches illustrating
sound and vision considerations on the roof terrace

builders could understand what they were building, she had arranged for them all to go and see a performance in the RSC's 'temporary' Courtyard Theatre. It was such a success that in 2010 she decided to do the same for the SWC team. 'Team building was even more important with the neuroscience project', she says. 'Putting people in a situation where they travel together, they get to know each other in a different way. We booked coaches, toured the site before the theatre was finished and then we went to see a show in the Courtyard, the theatre that Ian had already built. When the Royal Shakespeare Theatre was finished [in 2011] we took them again, the whole lot.'

Representative user Arifa Naeem had never previously been to Stratford-upon-Avon. 'Susie treated us to a really lovely lunch and the performance, it was a complete day out', she says. 'It did work as a team building exercise, but I find the whole team works very well anyway. Even if they don't agree it's dealt with in an adult, mature way. But yes, the trip did help. We talked a lot on the bus, and found out about people's lives, which you don't normally do at the meetings. I thoroughly enjoyed it.'

The strict timetable of the outline plan of work concentrated minds, and the time spent in workshops proved its worth, even for busy scientists. Chris Russell ticks off numerous points in the development of the design that emerged from the workshop process (these are all explored in more detail in the following chapters): 'How the building is arranged, with the BSU where it is; how that would be serviced; the novel satellite BSUs; accepting the concept of the two-storey lab spaces; whether the GCNU's space would be on multiple floors or not; how the building would be furnished. The double-height lab spaces were Ian's concept, but developed through the workshops. Certainly the louvres for the windows, and even the fact that there should be windows; the write-up and meeting room spaces; the location, layout and subdivisibility of the lecture theatre and the fact that it is not raked; the open amenity space at the back of the building, and the pocket park. Even things like where visitors would come in, and where lockers would be (and how that would be arranged versus the scientists putting their bags next to their desks); going to the loo, having a coffee, having a sandwich, as well as the workspace.'

By January 2011, the version of the SWC that had been designated as building study 23c had become the basis of Stage D, which would be costed, sent to the planning department at the London Borough of Camden, and presented to contractors bidding to build the laboratory.

4

INSIDE OUT

THE DESIGN EMERGES

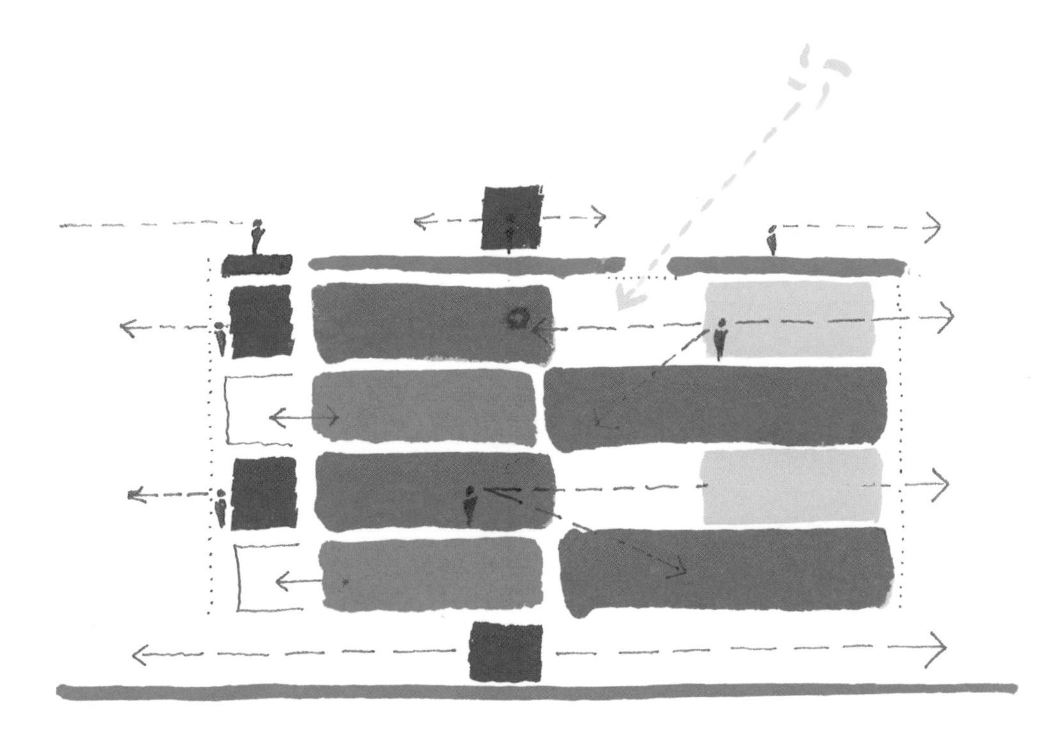

Ian Ritchie likes to make simple ink sketches to aid his thinking in the early stages of a design. As he began to analyse the brief for the SWC, he carefully arranged bright blobs in four different colours to represent the four principal types of space in the building. 'First, there is the laboratory itself', he says. 'That includes the GCNU: they are not offices, they are labs, but very different and understood to be different. Then there's the support BSU, which is fairly demanding technically but spatially also adaptable. The third area is write-up, which is either inside the lab or outside the lab. And the fourth area is for social interaction: break-out spaces where you can cook, have a coffee and so on. Those are the four distinct spatial typologies. But they are all tuneable, which I quite like.'

SPACES FOR SCIENCE

Apart from the 'bump' on the back, which is full of plant to service the building, the SWC is essentially a rectangular box, divided horizontally by flat concrete slabs supported on a regular grid of columns. How do you divide up the space within that box so that it meets everyone's needs? Flexibility is a crucial requirement, but some decisions are irrevocable. For example, the floor-to-ceiling height of

Ian Ritchie's ink sketch to aid thinking about the relationship of different spaces in the building

Services run overhead in exposed, prefabricated units suspended below the blue soffits

each level determines the vertical spacing of the concrete slabs that support the seven floors. Ritchie was determined to be generous. 'The aim was to give them unexpected space,' he says. 'One of the ideas that came out was the vertical dimension to the lab. So rather than the usual three-metre-high laboratory, we took the opposite approach.'

The convention in most modern buildings is that the wiring and other services run in a space between the ceiling and the soffit, or underside of the structural slab. Ritchie decide to gain height by dispensing with the ceiling, leaving the soffits exposed. Instead he decided to run the services – medical gases, water, electricity, data, sprinklers – in neat, prefabricated units suspended directly from the soffit. They are visible and accessible so that scientists can change the services delivered to their laboratory workspaces without disrupting the entire floor.

The concept took its inspiration from the interstitial floors in the Salk Institute, but is more economical in its use of space. Only the BSU, which needs its own plant and services, has an interstitial floor above, so that the services can be maintained or altered with minimal disturbance to the animals. The laboratory floors have been designed with vertical spacing of 4.2 m from floor to floor, including the depth of the concrete slab: this gives a working height of 3.7 m between the service runs, and 2.7 m below them, generous by comparison with both old and some newly-built laboratories.

Once he had started thinking vertically, Ritchie went further. A key part of the brief was that the building should optimise the interaction between researchers. 'The idea of interaction spaces, which have been a buzzword for the past 10 years in laboratory design, really originates from the difficulties when you have many floors', he says. 'Because you can't see the other floors, you don't go there. How do you break that? How do you create vertical connectivity? The word I like is connexity, an old English word. The traditional answer is that you do an atrium, but to my mind atria are a bit boring.'

Instead he and his colleagues developed the idea that each of the four laboratory areas (each in turn designed to accommodate three PIs with their research teams) would become a two-storey 'house', or maisonette. Potentially the three teams could each have their own tall, narrow 'townhouse', but ideally they would share the space. 'We ended up with double-height spaces in individual houses, rather than a generic atrium', he says. He proposed that the upper

RESEARCH TEAMS

The senior staff of a research institute typically consist of a director and a number of principal investigators, or PIs. PIs devise and direct their own research programmes, and spend much of their time raising or managing the necessary funding. Postdoctoral research assistants (postdocs) and students working towards their PhDs carry out most of the laboratory work. They are supported by technicians and administrative staff.

floor in each 'house' would be a mezzanine around a glass-walled, double-height space, crossed by two bridges and connected to the lower laboratory floor by an open staircase. The upper floor is designed to house the open-plan write-up area for students and postdocs, a small meeting room, separate offices for the three PIs and adaptable lab space. The mezzanine floor also has 4.2 m floor-to-floor height, which provides a potential working height of nearly 8 m in the double-height space.

Making a large hole in the floor slab presented the engineers with a whole new set of challenges. 'How thick does that slab need to be, how big can it be, what is the column spacing under it, how you get the air into the space?' says Jennifer DiMambro. 'We're effectively connecting spaces together from an ME point of view.' Chris Russell points out that if the technical challenges can be overcome, separating the write-up areas from the labs environmentally but not visually solves a number of different problems. 'If you have the write-up space in the lab, the problem is that the labs are much more highly serviced than office-type spaces in terms of air change rates, temperature control, humidity control and so on. So by taking the write-up out of the labs, it is much more efficiently ventilated. And you can open the windows, which you wouldn't want to do in the labs, and you can eat and drink at your desk, whereas in the lab you are definitely not supposed to do that.'

WRITE-UP AREAS
Most contemporary researchers spend up to 50 per cent of their time at their computers, analysing data or writing up their results. Whereas once a scientist might do this work on the laboratory bench, health and safety considerations now dictate that separate write-up areas need to be provided. These may take the form of individual offices for more senior staff, or open-plan carrels for graduate students and postdocs.

Heated discussions arose in the workshops about where people preferred to do their desk-based work. Some were used to doing it inside the laboratories themselves: others wanted them separate. 'You can tell scientists, "If you're working on a lab bench, you shouldn't have your write-up in the same place"', says Jennifer DiMambro. '"You shouldn't be taking a cup of tea onto your lab bench – there are really good energy and health and safety reasons for separating it." And they'll go, "This is what I've always done, this is what I want." They have a very different perception of what a good building is.'

'We wanted to separate it for lots of reasons', says Russell. 'But you then have the potential for losing contact with the labs. So the double-height spaces came from the desire to keep a fairly close relationship between the PIs and the write-up, and looking down into and across the lab, so you always have a visual connection between the write-up and the lab. That's a relationship that I haven't

Each lab is on two levels, connected visually and physically via by a glass-walled mini-atrium

seen anywhere else. And having exposed concrete soffits, and services providing as much height as we can, including the double height space – we hope that a number of scientists might take advantage of that in their research. How often do scientists get the opportunity to use 8 m of height? You could have a vertical experiment!'

The exception to this arrangement is the Gatsby Computational Neuroscience Unit, almost the only occupant of the new centre that was identified before any design work began. The computational neuroscientists have very different needs from their experimental colleagues. They work with computers, pencils and paper or whiteboards. They have no need for biological material or chemical reagents, so the servicing of the spaces is much simpler (although they do need uninterrupted access to a powerful source of data processing).

The GCNU's director, Peter Dayan, was very concerned that his group should continue to work effectively in the new building. 'I start from the fact that this unit is not conventional', he says. 'We have two groups of people who don't normally work together: machine learning people and theoretical neuroscientists. So the first thing in my mind is to make sure the things we have that work don't break in the new building. How we talk to each other across our own interdisciplinary divide, how we enforce that, what are the benefits – how that will work in a new building is hard to assess.'

To begin with Dayan had his own idea about where in the building it should be located. 'One of the issues when they were designing it was that we need light', he says. 'The obvious place for us to live is on the roof. We were offered two sites in the building: one on the roof, the other in the centre of the building. We hoped to be on a single floor. But the cost of that was that we would look like we were ignoring everything else that was going on in the building.' Ian Ritchie says, 'I personally didn't give in to Peter wanting to be all on one level. Nobody would ever have seen them. So he is inconvenienced by being on three floors. I was under clear instructions that the theoreticians must meet with the other people. So he got the three-storey house, that was the model.'

The next question that arose was over the nature of the working spaces allocated to the GCNU researchers. Elsewhere the write-up areas were mostly open plan. 'I talked quite a bit to the postdocs here, and their strong view is that they wanted to have offices', says Dayan. 'They are really labs too, because we don't have any other labs. All our work is in these labs. If you have open-plan write-up areas, you have to have break-out spaces where you can have discussions so that

you don't disturb people.' On this point Dayan won his case. 'We have maybe two or three single-person postdoc offices, and others are for two, three or four, so most of the students will be in those', he says. 'But not a 30 person office, which the write-up area is. That was one of our bottom lines. People feel that was critical to the way they work. So Ian thinks of us as monks who want to have our little monastic cells.'

Within his three-storey house, set in the centre of the building on levels 2-4, Dayan and his colleagues have a number of these 'monastic cells' conducive to concentrated thought – either with a window opening onto Howland Street, or facing into the central light well. The offices cluster around a central double-height seminar room with its own tea-making area. 'We were most concerned to replicate

Glass-walled offices in the GCNU provide plenty of surfaces to write on

our seminar room', says Dayan. 'For us it is the heart of our unit. We meet collectively and have tea every day, and someone gives a short talk.'

Gordon Talbot remembers that there was creative tension throughout the design process between the GCNU's isolationist tendencies, and the desire of the Gatbsy Foundation to promote interaction between the theorists and the experimentalists. 'Sarah [Caddick] made some very interesting observations about the GCNU in the centre of the building: that the labs either side should be porous', he says. 'So that Peter is in a sandwich between the labs, but there is a corridor through the GCNU.' The corridor passes along a gallery on the upper level of the seminar room with views and windows opening into it. Dayan talked jokingly about building a moat around the unit and asked for doors to act as drawbridges that they could pull up when they wanted quiet moments to think together as a group. 'Sarah said "Those doors must always be open unless somebody makes a decision to close them"', says Talbot. 'So we've put these hold-open devices on the doors so that they can always be open, but they can be periodically released and closed.'

ALL ABOUT ANIMALS

A neuroscience laboratory with a focus on neural circuits and behaviour collects most of its data from experiments on living laboratory animals. The wellbeing of those animals is of paramount importance: for their own sake, to meet the stringent legislative requirements surrounding the use of animals in research, and because if animals are distressed or unwell they will be unsuitable as experimental subjects.

For that reason, an early priority of the SWC design team was to get the BSU (biological services unit) right. How big should it be? Where in the building should it go? How should it be arranged? How should it be serviced? These were among the questions addressed during the first of the scientific advisers' workshops. An even earlier decision concerned the species of animals that would be housed there. The funders and advisers of the SWC decided from the start that the majority of experiments would use mice. Because of the need to remain adaptable in the future, the BSU would be designed so that it could be adapted to accommodate other small mammals, such as rats or ferrets, or other classes of vertebrate such as fish.

WELFARE IN THE LAB

Under the Animals (Scientific Procedures) act of 1986 (amended to take account of an EU directive of 2010), every experiment must be licensed by the Home Office, every scientist who handles animals must have a licence, and the premises where experiments take place must also be licensed. The welfare conditions governing the issuing of these licences are among the most rigorous in the world.

A high-level gallery provides a viewpoint on the GCNU seminar room

'The deal that was struck very early on was that we didn't know what kind of research would be done', says John O'Keefe. That decision would be left to the Director, yet to be appointed, and would depend on the choice of group leaders, yet to be made. But it was unlikely to include insects such as fruit flies, commonly used in behavioural research.

'The SWC will concentrate on rodents and probably fish', says O'Keefe. 'The problem in this game is to pick a type of behaviour or a type of sensory process and then to try to pick the organism that is farthest down the evolutionary scale that instantiates all the important problems but is simplest to work on. It's very clear that some people feel that rodents might be too complicated. The zebrafish larva has something like 100,000 neurons in its brain [versus 4 million in the mouse and around 20 billion in the human]. They have a lot of behaviours so they are very attractive. You can see the whole brain, they are transparent.'

Any innovation in the design of animal facilities for the SWC would have to gain the approval of the Home Office. In addition to regular three-monthly meetings with the Home Office inspectors, the team convened a BSU working group that would explore the issues in detail and report back to the scientific advisers' workshops. It included Adrian Deeny, the Director of Biological Services at UCL and Barry Warburton, one of the managers involved in the day-to-day running of a BSU at UCL. 'The group allowed us to talk very freely but also in a very focused way', says Deeny. 'Setting it up was a very good move. There are times when the BSU is discussed as part of the bigger process of development and important things get overlooked. The BSU is a fundamental and very key part of any research establishment, and although people might not like to advertise that it's there, without it nothing happens. It was a very good initiative, everybody had a vested interest in making it work properly and there were lots of new things that we were able to put in.'

MOSTLY MICE
Laboratory mice are small, and efficient to feed and house; they are easy to keep and they breed rapidly. Their genome sequence is known, and studies are beginning to understand how genetic variants affect behaviour. As mammals their fundamental biology is identical to that of humans. They learn to respond to changes in their environment. The lab mouse is therefore a popular choice as an object of study for neuroscientists interested in behaviour.

The neuroscientists that the design team had met in the course of their lab visits were of one mind in preferring to have laboratory animals housed close to their laboratories. Traditionally, institutions have housed all their animals in the same space, usually in the basement or at the top of the building. At the Salk Institute, which in many other respects was a model for the SWC, the animals were in a separate building from the labs. This arrangement works well for the animal care staff, who are able to ensure that the animals are kept in optimum conditions, but the scientists themselves don't like it. It means either that they have to work in two separate locations, or that animals have to be transported backwards and forwards between the lab and the BSU, causing stress and raising the risk of spreading infections.

Richard Morris had been thinking about how to avoid this from the beginning. 'Seeing what the set-up was at the Salk, and some other labs we visited, was what led me to propose something that I'd done in my Edinburgh laboratory back in 2001', he says. 'I hadn't initially twigged it could be good for all neuroscience, not just behavioural neuroscience. That was to have a main animal house, and then separate satellite animal houses. So I started to push for that on the design committee. Could we begin to imagine that part of the animal house is up near the labs with some rooms around it for experiments?' John O'Keefe was enthusiastic. 'In most places, if you want to have an animal that's running in a maze, you go all the way across the building, get the animal, put it in a box and carry it back again', he says. 'For some of these animals, that could be stressful. It makes the research not very easy to do.'

At his own lab in UCL he had been able to keep a small group of animals close to the testing laboratories, and he was keen to see that arrangement designed into the new building. Sarah Caddick was very supportive. 'I really wanted to see animal facilities for every lab on every floor', she says. 'It's better for the animals. The way science is going now, there are going to be more longitudinal experiments done in awake, behaving animals. You can't be sending your postdocs off to a BSU in the basement all the time. For behavioural studies you want the least disruption possible for the animals.'

Addressing these conflicting interests was one of the first challenges that the design team undertook, with advice from David Kelly and regular consultation with the Home Office. 'The BSU managers want the exact opposite of what the scientists want', says Chris Russell. 'They want the minimum number of people to interfere with their facility, because the more people who come in, the greater the risk to the animals' health. So the BSU managers want to isolate the facility from everybody, but the scientists need access to the animals for their work.'

The answer, arrived at after considering many different arrangements, was to link the main BSU in the sub-basement with 'satellite' holding areas adjacent to each of the main lab spaces. 'It's taken us back to the situation 30 or 40 years ago where people had animal facilities next to their labs', says Kelly. 'Through a combination of Home Office and health and safety regulations, it's become

customary to build specialised animal facilities. Most people who run animal facilities prefer that. But with this particular type of work it is important to have animals close to the laboratories, because you don't want to be moving them up and down in lifts.'

What is novel about the design solution, which was approved in principle by the Home Office at an early stage, is the way that the satellites have been linked with the main BSU. 'We've taken the environmental controls in the BSU and continued them out into satellites', says Kelly. 'So we are not compromising at all on the facilities, and I think it will be a good way forward.' The scientific advisers were extremely enthusiastic about the proposal. The main BSU in the basement would be linked with the satellite holding areas by a lift opening into a service corridor on each of the lab floors, and by dedicated ventilation and drainage systems. The complexity of such an arrangement emerged during the lab visits made by the design team. 'One other building we saw attempted it', says Ian Ritchie. 'In their labs they had satellites, but they had no floor drainage. They'd forgotten things you needed, so they couldn't use them as satellites. When we saw that, we grasped the rigour of understanding exactly what one of these would require.'

'Scientists have relaxed animals very close to the lab – an animal will be more comfortable moving two metres than being brought up two or three floors in a lift', says Russell. 'You can move them up a few days early, get them used to their new surroundings, then move them a very short distance into the lab. The importance of the animals being calm, relaxed and happy for the scientific results was something we learned a lot about from the lab visits.'

Lifts connect the main and satellite BSUs

A service corridor runs behind the labs, also with exposed services

John O'Keefe is very pleased. 'People are quite envious', he says. 'Others have not been able to do that.' Controlling the temperature, humidity, ventilation and other environmental factors in the satellite areas to the same standard as in the main BSU was a challenge for the engineers, but not an impossible one. 'The Home Office insisted on all sorts of things that made the building much more difficult to design', says O'Keefe. 'For example, we have three completely different ventilation systems, one for the write-up areas where the humans will be, one for the labs and one for the animals.'

'Each satellite area is basically a box within a box' says engineer Jennifer DiMambro. 'You've got the normal lab services running through to the lab, then you've got a ceiling, so that it's a box on its own. And then you've got special services that just service that box, and another ceiling under that.'

'The satellites are an amazing thing to have', says Arifa Naeem. 'They are right next to the lab, you don't have to go down to the main BSU. I don't think there's anything anywhere else like this. This is a new thing, being asked how you want your lab to be designed. We are used to being given a space and told to make the best of it. It's a privilege and a blessing.'

With thousands of mice accommodated in the main BSU at any time, it is essential to avoid infection. In recent years the need has become even more acute because of a change in the way animals are acquired and kept. While once all animals had been bred off site by a commercial breeder, many laboratories now use genetic manipulation to breed mice that have special characteristics. 'Instead of being holding rooms, these rooms have become breeding rooms', says Deeny. 'In the old days you used to get infections all the time. But because you were only holding animals for experiments, you did not have so many potential long-term problems. When you've got breeding animals they are there all the time, so an infection has a real impact on that colony.'

One of the innovations adopted for the SWC to protect the mice from infection is the individually ventilated cage, a technology David Kelly first saw more than 30 years ago in the USA. 'They live in their own little environment', he says. 'You have a rack of cages, each of those is completely enclosed. Air is blown into each individual cage, and air is exhausted from each individual cage. They are very expensive but they overcome a lot of problems. They not only provide the animals with better protection from disease but also the animal care staff from allergy, which is a big problem in the BSU.' They also reduce the demand for ventilation of the rooms themselves, and so have little impact on overall running costs.

The only concern was that the new system might affect the behaviour of the animals, and so make it difficult to compare the results of experiments with other work on animals housed in conventional cages. 'The basis of all science is that you keep everything on a level playing field', says Kelly. 'Wherever I've had these concerns I have raised them with John O'Keefe and the Scientific Advisory Group. Early on we came to the conclusion that we should be planning for the scientist to be able to use either individually ventilated cages or conventional caging. I think that was a smart move.'

An enormous amount of thought has gone into controlling not just the air quality, temperature and microbiological environment, but sound and light for a species that is nocturnal and disturbed by vibration or loud noises. 'One of the things none of us had thought about [before the lab visits] was fire alarms in animal behaviour units or in the BSU', says Sarah Caddick. 'The frequency of the sound, or how it disturbs the animals, can ruin the experiment. At Duke University the BSU people had figured this out: there's a type of alarm you can get that emits only light and not sound. That's one example where we thought OK, this is good.'

It is probably accurate to say that more time and expense have been lavished on making sure that the animals are comfortable and healthy than on the design of spaces, equipment and services for the people in the SWC. However, the project team were anxious to make sure that the animal care staff also had a pleasant working environment. 'We said, "When you are working in the BSU it would be

Skylights allow natural light into the GCNU seminar room, where tea and cake is a daily event

nice to have light"', says Naeem. 'Mostly BSUs are in the basement or on the top floor with blinds down, but it can get really depressing working in a space without any daylight.' Sarah Caddick and Susie Sainsbury were enthusiastic advocates of this view. 'Typically they get a little staff room but it's still part of a warren of rooms in the basement,' says Caddick. 'In the SWC the staff room is in the middle of the warren, but it's double height, and glass-roofed. So they not only have daylight, they have the same generous room height that the scientists and administrators have in the rest of the building.'

'I think they've worked really hard to make conditions for the animal care staff very good', says Deeny. 'The technology they are putting in to introduce natural daylight into the sub-basement is fantastic. And if you can't have open windows, I think we've got the next best thing.'

A SITE OF INTERACTION

The social dimension to the design of the whole centre is at the heart of Ian Ritchie's thinking on the project. Collaborative working is both having time together and time apart: communal discussion and private reflection, and these require different and accidental spaces. 'I also thought about what drama or incident one could create spatially, that would be attractive to them as a place to have conversations', he says.

Recalling his discussions with Richard Axel, he goes on, 'The real ideas in research come up in conversations, and they happen anywhere.' Promoting interaction is partly a matter of creating specific 'loose yet focal spaces', such as coffee and tea points and meeting rooms, but also of thinking about how people move round the building and collide with one another. 'The idea of passive interaction, when you meet somebody in the kitchen and talk about things, is very helpful', says Talbot. 'A lot of science buildings have headed off down the road of labelling rooms 'interaction space', and the creative industries have gone way overboard with creating focal spaces. It's fun, it's what they do, but it underlines the point that a big, open-plan lab or office is a bit soulless. You're a bit stuck when you go and talk to somebody and everybody's watching the fact that you're talking and not doing.'

The lab visits gave the team many opportunities to assess how successfully other buildings had fostered productive interaction.

INTERACTION SPACES
Tea rooms or common rooms were often features of laboratories built in the mid-20th century, and the canteen at the LMB in Cambridge where Francis Crick held court has passed into legend. More recently a kettle in the corner of a seminar room was the most that was on offer in many research departments. As scientific research transcends traditional disciplines, buildings encourage informal interaction with small break-out spaces as well as central cafeterias.

Meeting rooms on the top floor enjoy views and access to the roof terrace

The main staircase ('Ariadne's thread'), painted in gold, is the only exception to the blue and white colour palette

But John O'Keefe was initially slightly sceptical that creative interaction could be engineered. 'The questions in the scientific advisers' workshops would be "Where are people going to bump into each other?"', he says. 'That is a very dominant view among people who think about how science works. I do believe that people get ideas from each other, but equally there's the idea that somebody goes away and sits in their monkish cell and thinks very hard, and to be honest I think that's much more the British tradition.'

However, no one wanted the arrangement that the visiting team saw in one laboratory: a long corridor of doors with names on, and no windows. 'You can't say "Hi Sarah, Hi Bob", and they might give you a wave to come in', says Talbot. 'You don't want to knock on the door, so where else do you meet? Sarah and Susie said "Coffee is everything" and Peter Dayan said "Coffee is definitely everything, and by the way don't forget tea." But then there are other natural spaces, that's what we were striving for – it could be the stairs, it could be the corridor. So we just widened out the circulation spaces to try and create informal meeting places in front of the stairs. It was a very simple move. But hopefully it isn't contrived and they are open spaces and they'll work.'

CIRCULATION SPACE
In architectural terms, circulation spaces are the areas through which occupants of a building pass on their way to or from the rooms they occupy for longer periods. They include staircases, corridors, foyers or atria. In a well-designed building they offer opportunities for unplanned encounters that increase the social coherence of the community of occupants.

Where soffits are concealed in meeting rooms, lighting continues the blue and white theme

For more formally arranged meetings, there are a variety of options. Each of the laboratories has a small meeting room at the end of the building, with huge windows and views out into the street. At the other end of the scale there is a 170-seat adaptable lecture theatre. Finding the best place to put this was one of the challenges that was resolved through debate in the workshops. 'For a while it was at level 5 [on the roof], being the place with the best views and a bigger footprint not constrained by structure' says Russell. 'But in the end the scientists and the funders were concerned about getting everybody that might be coming from the outside up to level 5 quickly. We were told that scientists don't show up for a lecture until one minute before it starts. They were much happier having the lecture theatre on the ground floor.'

The lecture theatre can be divided into two or three smaller seminar rooms using horizontally folding walls; Russell thinks it works really well. 'Previously it didn't have as much potential to be a public part of the building, whereas now it could be', he says. 'You come into reception, and immediately adjacent to reception you've got the lecture theatre and support spaces that serve semi-public events.'

The main lecture theatre on the ground floor is adaptable, with folding partitions to divide it into smaller spaces

It is even possible to retract all the walls and make it a single exhibition space. Keeping this adaptability was why the architects decided to go for a flat-floor lecture theatre, with stacking chairs and moveable walls, rather than a raked and formal auditorium.

Meanwhile Susie Sainsbury was concerned that the 120 staff in the building should have a place to eat together. Some pointed out that Bloomsbury had one of the densest concentrations of sandwich shops and restaurants in London, but she was determined. 'By having a canteen producing lunch every day, everyone sits down and talks about each other's projects', she says. 'We're trying to use that eating environment to produce a sense of cohesion among a mass of people who may not have anything linking each other. It was a feeling that Peter Dayan's team won't have met the technicians that look after the animals, helping them to have a forum where they can bump into each other. And there are catering companies that can do very good, healthy food cheaply.'

MEETING ROOMS

As well as experimentation, exchange of information is crucial to the practice of science. Scientists regularly present their information to each other through illustrated talks, whether it is a student presenting a result to the rest of the group, or a distinguished visiting speaker lecturing to the whole institute. Seminar rooms of different sizes are essential, and a multi-purpose lecture theatre is valuable even in a non-teaching institution.

The top-floor brasserie doubles as extra meeting space

A brasserie was designed to occupy the desirable space on level 5, with access to the roof garden and views across London. This space is also adaptable and can be reconfigured to create two spaces – a 30-seat meeting room, and a smaller brasserie.

FIRST SERVICE

If the concrete frame of a building is its skeleton, then the services are its circulation and nervous system. And just as evolution has developed the two in parallel in the human body, the architects and engineers worked together to ensure that structure and services worked in harmony. It made no sense at all to Ian Ritchie to make a beautiful building and then knock holes in it to run cables or pipes to where they needed to be. Having grasped the principles, the architects then led the services design arrangements and with the engineers worked very hard to ensure solutions that were both practical and aesthetically satisfying.

That meant thinking about how the building would be constructed, rather than how it would look when complete. 'We had a whole series of studies looking at where the ducts run', says Russell, 'including thinking about whether or not to have interstitial floors. We have one for the BSU, but otherwise we have what we called 'endostitial' servicing – the lab ducts are on the outside of the building, at the back. They drop down from the air handlers on the roof, so you don't have large ventilation risers going up the inside of the building breaking up the space. With the BSU, it was about keeping the BSU plant - filters and air handlers – on a separate floor to minimise disturbance and facilitate maintenance.'

As well as the plant, all the services distribution for the BSU is on the floor above, separated by a concrete slab. 'Designing for the animals is much more controlled than the rest of the lab', says DiMambro. 'It's all subject to Home Office approval. It has to provide the right level of resilience, the right health protection for the animals, stop the ingress of disease or the spread of disease. It's something that can be validated and meet the Home Office criteria. For example, acoustics is quite a critical feature of the design. If the inspectors come in and it's too noisy they won't open the facility. And if you fail the Home Office criteria you just can't use the building. There's a lot of rigour. And the same system comes up the building to the satellite holding areas.'

HVAC SYSTEMS

Large, modern office buildings are kept comfortable by heating, ventilating and air conditioning (HVAC) systems. These can be noisy, their plant and ducts occupy a substantial amount space, and they consume energy. In a laboratory, with potentially noxious fumes and a lot of electrical equipment, their operation has to be precisely tailored to provide clean air and stable temperatures. An HVAC system consists of central heating and cooling plant, a distribution system of ducts and fans, and intake, exhaust and heat recovery units.

In contrast, in the laboratory floors the services run within the space. The key to keeping the exposed services tidy was to have first built mock-ups of a typical lab structural bay, and then to have had services constructed off-site as prefabricated modules that could be slotted together as the building went up. 'Prefabrication is the way the industry is moving', says DiMambro. 'You have much more control over quality and safety on the floor of a factory. For the architecture, the fact that the services are exposed, the prefabrication gives you much better quality. It's a real step-change.'

About the principle of having exposed services, she was in two minds, but recognised that it was part of the architects' design vision for the building. 'You can argue the case either way', she says. 'Ian's vision is that the scientists can see where everything goes. I think it's more about intelligent servicing design and making sure everything is available rather than having a ceiling or not having a ceiling.' The architects, for their part, had seen so many untidy services installations, 'a throw it up and cover it up' approach to lab spaces coupled with poorly designed access, that they were sure that the future adaptability would be better served with easily accessible services. There are valves and connection points at regular intervals, so that once the space is fitted out the facilities manager can get on a platform and make the connection, and then change it easily as required.

A clear point of difference between the engineers and the architects, as mentioned in an earlier chapter, was the question of what goes on the roof. 'Ian sent me an email that said, "This building is all about services – if you were designing the building, tell me how you would design it"', says Di Mambro. 'But then he said "Don't put anything on the roof." Generally in a laboratory you put everything on the roof – if you put it in the basement you've got to get air down from the roof. Polluting things like generators usually go on roofs. At one point there was a slide that he inserted into my presentation without me knowing. I was doing this presentation to UCL and suddenly this slide popped up – it was a garden, a whole lawn all over the roof. Ever since then I've been this villain who's destroyed their vision and put plant on the roof. We spent a lot of time on that.'

The outcome was that the chillers, generators and air handling units were raised up on high tables above the roof, so that there was still room for a garden, and both sides felt vindicated. 'The reason Ian said to Jennifer that you can have plant anywhere except on the roof was because the roof was for people', says Chris Russell. 'It has the best views any lab building in London can offer. The brasserie is underneath one plant table, and the admin offices are under the other. A link connects the two sides, with a north-facing roof garden and a south-facing terrace.'

In the middle of a city with mains electricity, why did they need generators at all? 'The building is very resilient', says DiMambro. 'Because there is a BSU in the basement, there is a concern to protect the welfare of the animals – more than there is for humans. Under Home Office regulations you need to be able to support the facility for five days without power. Of course, if central London doesn't have power for five days, we've got a more serious problem than the animals in the basement!' To increase the resilience of the mains electrical supply, in addition to the generators, it is divided into two branches. 'We were going to bring two separate power supplies into the network,' says DiMambro, 'but we would then have had to reinforce the local area which would have cost £1m. So we have a single supply into the building, split into two separate systems. There are two networks, and two generators so that if one fails you have another. It led to a lot of debate once the contractors came in and looked at the costs – one of the first things they said was "We'll get rid of the generators." But we fought hard for them because it enables the science. There was this idea that you took one or other out, but we said, "We've got to look at how the building will be used in the future."' UK Power Networks, the company that supplies electricity to London and the South East, has since begun paying the SWC to keep the generators on standby in case of dips in the power supply in the area.

The stability of the power supply is not just about keeping the animals safe and comfortable. 'There's lot of data that's produced', says Christian Allison. 'The power to keep data stored on the servers if the building's supply goes down is very important. People could lose years of research.'

The size of the data centre that would be needed, equipped with high-capacity servers, was another topic of discussion. On the one hand the sky is the limit in terms of potential future demand, but on the other in the contemporary data processing environment you don't necessarily need to have all your storage on site. 'There's a space issue and a power issue', says Peter Dayan, whose GCNU team have a significant interest in capacity for data processing and storage. 'Nowadays they're building data centres in places like Iceland to get geothermal power and cooling. But you don't need to have all your computers in your building. UCL is building a data

DATA CENTRE

Microscopes, genetic sequencers and physiological recorders generate vast quantities of digital data, as do computational studies of brain function. To support the storage, processing and distribution of these data, laboratories need access to one or more data centres. A data centre is a collection of high-capacity networked servers. Data centres need not be on site, although most laboratories have a moderate data processing and storage facility within the building.

centre just inside the M25. The other thing is that data storage is so much more efficient.' These were the factors that the Arup engineers had to consider when deciding how much power to allocate to computing and cooling.

'In central London, do you want to have a massive data centre using up prime real estate?' asks DiMambro, 'or do you use cloud-based services, or outsource it?' The final decision was to allocate 100 m^2 to the SWC's data centre, a decision taken partly after discussion with UCL's IT managers. 'That's quite a good example of interaction', she says. 'None of them wants to commit to where data is going, because it grows so exponentially. We had good discussions about how they see the subject moving forward. There are all these terabytes of data everywhere, but one of them said "I only have this much budget, so I can only install this much computing year on year." Because the demands change so rapidly and the technology changes so rapidly, we had to find a balance between putting so much power in the building, and making what we thought was a good estimation for the first three to five years.' Peter Dayan is satisfied with the outcome. 'We have what we think will be enough for the next few years', he says, 'and will buy further services as and when we need them.'

BUILDING FOR THE FUTURE
BREEAM was devised in the late 1980s at the Building Research Establishment in Watford, England as a set of standards for measuring sustainability in building. The BREEAM New Construction system of certification was introduced in 2011 and covers all new building in the UK. BREEAM certification is a requirement placed through the planning system and/or government agencies.

The engineers had to counterbalance the need to keep the building going at all costs with the need to be as efficient as possible in terms of consumption of energy and materials. The architects, the clients – and the planning authority, the London Borough of Camden – were looking for a BREEAM assessment of 'excellent', a benchmark indicating a high level of sustainability.

This would be a challenge for this type of laboratory building with such a high demand for services, and some technologically innovative solutions have been integrated. 'The thing that's really innovative from the services point of view is a demand-based ventilation control system', says DiMambro. 'It comes from the USA, and has only just been approved for use in the UK.' It is a technology that Arup have been keeping an eye on for a number of years, knowing that the company, Aircuity, was trying to get EU certification.

'It's basically a way of controlling the ventilation, and so the energy consumption, in the lab', says DiMambro. 'It senses contaminants in the air. Normally you constantly put a huge amount of air through a lab to make it safe. You have to work on the worst-case scenario: someone's working with something nasty, or someone's spilt something. But the reality is if you walk round a lab it's very rarely fully occupied, and a lot of the time no one's doing anything smelly or dangerous. So there's a huge amount of air going through the building 24/7 when there's no justification for it.' The Aircuity system has sensors in the ducts that sniff packets of air and take them back to a monitoring suite. It measures the total concentration of volatile organic compounds (VOCs), as well as particulates such as soot or liquid aerosols. 'If it spots any changes it ramps the air up', says DiMambro. 'It can reduce your running costs substantially.'

The SWC is the first new build in the UK to install the system, though DiMambro has been to visit a smaller system installed in a newly refurbished laboratory for the Medical Research Council. 'I'm hoping it will be a success story', she says. She estimates that it will save at least 200,000 kg of carbon per year. Together with a combined heat and power (CHP) engine in the basement, itself estimated to save 500,000 kg of carbon a year, they were able to present an application to the planning authority that complied with its 'lean, clean and green' policy. 'In a central London site it is difficult to do renewables', she says. 'We did look at where we could put photovoltaics [solar panels], which we thought would save about 5000 kg of carbon per year. But as an alternative we asked Camden to consider this [Aircuity] as the renewable – it was not financially possible to do both – and they agreed.' Ian Ritchie remembers the moment. 'That was huge', he says. 'It made everybody happy.'

A similar technology to control lighting according to demand provoked more discussion. 'I've been in buildings where people sit there and wave their hands in the air,' says John O'Keefe. 'And eventually I've said, "Can somebody tell me if I'm supposed to be engaging in this little dance?" And it's because the lights will go off if there's no sign of movement. So there are problems. If you are doing animal work you need to be able to control the lighting and have control over the actual programmes themselves. They have to be simple enough. So we've put a lot of effort into making sure that's the case.' DiMambro was sympathetic: daylight sensing is installed in Arup's own offices. 'I hate lights that switch off,' she says, 'but we always do it because it saves energy.' The lighting system at SWC will enable the scientists to override the automatic settings when they need to keep light levels constant in the labs, using a simple four-setting control panel. In the BSU staff and circulation areas 10 m below ground, the design team has taken advantage of recent advances in 'biodynamic' lighting, which subtly changes the spectrum of wavelengths to mimic natural daylight and harmonise with the body's natural daily rhythms.

GOOD VIBRATIONS

A building in the middle of a busy city can't help but shake. Two underground lines (the Northern and the Victoria) run on converging routes that take them each within less than 100 m of the SWC's site on Howland Street. There are roads on three sides, and the busy, bus-filled Tottenham Court Road only two blocks away. And the building creates its own vibrations – fans, generators, fridges, as well as footfalls on the floors and staircases. 'All structures have a natural frequency', says Ritchie. Yet somehow much of that vibration had to be designed away. 'For the nature of the equipment being used, we had to achieve quite stringent levels

The columns in the lecture theatre help to engineer vibration out of a building surrounded by busy streets

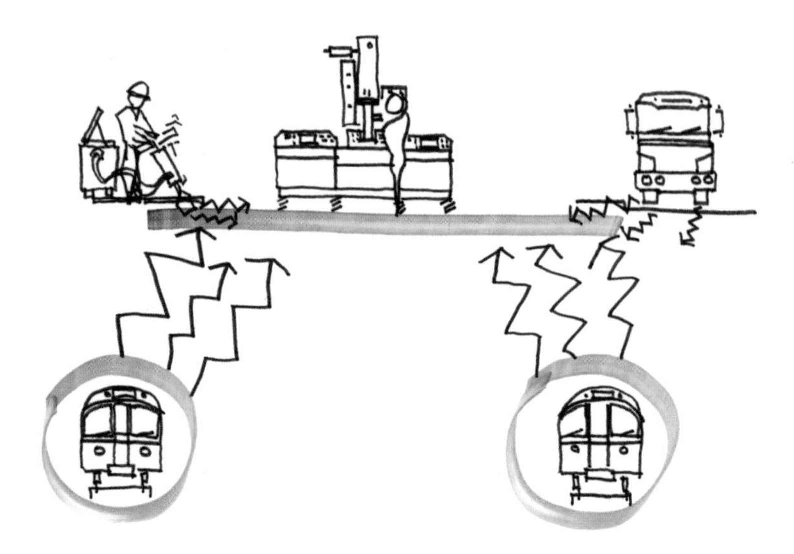

of limited vibration', he says. 'That is why we have relatively short spans [between the supporting columns], and why there are two columns in the lecture theatre, because right above it is a lab.' The big CHP engine in the basement also sits on its own isolated slab in an acoustic enclosure, as do the two generators on the roof, so that they do not shake the rest of the building.

But what about the other floors? Controlling vibration was seen as so important to the performance of delicate microscopes and recording devices that the project team scheduled a scientific advisers' workshop exclusively on that topic. The engineers put in a lot of time on research before that meeting. 'We tested laboratories around the UCL campus to get feedback from them about what they think the performance of their space actually is, as opposed to what they think it should be', says DiMambro. 'We were trying to get that element of reality between setting criteria for a space and throwing money at it to make it work.'

To achieve the level of vibration control the scientists believed they needed through the structure itself would indeed be costly, involving the installation of viscoelastic coupling dampers (VCDs) into the floor slabs to isolate the instruments from the building's resonance. Sarah Caddick had a suspicion that even if the scientists were presented with the data on the performance of their floors, they would still be unconvinced that it was good enough. 'Vibration is a big issue for

Architect Gordon Talbot's sketch illustrating the challenge to delicate scientific work from construction, traffic and two underground lines

the kind of experiments they will do', she says, 'so I wanted to bring in the scientists to discuss all of these big decisions we were going to have to make.'

What everyone had noticed during the visits to the labs was that scientists using sensitive equipment mounted it on vibration isolation tables, known as air tables as they use compressed air to act as a cushion. Chris Russell, who was tuned into this debate, says, 'It came down to a question at a scientific advisers' workshop: "If somebody told you that you had a space that met the vibration performance criteria that you needed for your research, would you still buy an air table that cost £10,000?" And the scientists said, "Yes, even if the VCD slab is working, I'm going to put an air table on it anyway." So everyone agreed that it was not worth spending additional money on giving the structure a higher vibration performance if the scientists wouldn't trust it.' The cost of the air tables was trivial compared with engineering the building, and it meant that on the laboratory floors, the structural level of vibration control could be less stringent, as cost-effective air tables would make up the difference between what the building could do and the equipment's vibration performance requirements.

ALL SHOOK UP
Different parts of a building oscillate in response to vibration transmitted from inside or outside the building. Controlling vibration is important for the safety of the building – a particular consideration in earthquake zones – for the comfort of the occupants, and for the operation of instruments within the building. Vibration control is a complex art, requiring a combination of stiffening and damping of the components of the concrete frame.

'We did a lot of work on that and it's quite a key piece of work', says DiMambro. 'It's a good example of how to optimise the building for the scientists without throwing money at it', adds Christian Allison. 'It was slightly counter-intuitive to us initially. We thought we were trying to achieve the brief – you start doing engineering gymnastics to try and achieve it, and then you realise you'll end up using more material when it's not strictly necessary. But we had to take people on a journey to get there.' 'Sometimes it's the psychology, not the function', says Caddick, who predicted this outcome. 'Though to be fair, one of the reasons the scientists were going to buy the tables anyway is that the tables come with lots of little holes, and they can screw all their equipment into them.'

Low-frequency vibration was not the only issue. The animals in the BSU are sensitive to a wide range of vibration: not just the rumble of a passing underground train, but high-frequency sounds including ultrasound. Laboratory rodents can hear sounds up to 80 kHz in frequency: humans can detect sound only to 20 kHz. As Chris Russell explains, nothing was left to chance. 'Nobody knew if LED lighting gives off an ultrasound signal because nobody cared – it's outside the human hearing range', he says. But 'suck it and see' was not really an option. Before any lights were ordered, the team arranged to have them tested with sensitive sound equipment to ensure that they would not interfere with the research. 'We also had discussions with experts about hand-wash basins in critical areas', Russell adds. 'We learned that water hitting a stainless steel basin causes all sorts of ultrasound signals: porcelain is better, but getting rid of the basin and using gloves and antibacterial gel is even better.'

ADAPTABLE AND FUTURE-PROOF

Throughout the design process everyone had to remember that they were designing not for themselves, but for scientists yet to be recruited; and not just for now, but for half a century to come. 'The demand on us as designers was to provide adaptability', says Ritchie. 'Adaptability was fundamental and ran through all the workshops, and we fought to keep it throughout construction.'

The need to adapt comes partly from the likelihood of technical change, or the turnover of research teams. But it is also partly to do with the nature of the science. 'One of the big surprises early on', says Gordon Talbot, 'was that whereas many have the idea that scientists sit at benches in white coats and mix things and look through and at instruments, when we went to see neuroscientists it was a different world. They were in workshops with wires and cables and boxes and bits of strut and things for hanging things. And we found that they had actually built a lot of this equipment themselves. What we had to do was create the backbone for all of that to happen: a bit of light, places to hang things, places to build things, a lot of room to get things in and out of the building.'

On the laboratory visits they had discovered that many scientists were hampered in what they could do by what Talbot calls a 'central services' attitude to adaptation. 'If they wanted to add a bit of equipment someone had to come and take the suspended ceiling down and add some pipes. Three months after that someone else would come and put the taps on. Three months after that the ceiling would be put back, and three months after that they'd get a bill for £8000. We wanted to short-circuit all of that by making the services available so that they had something to hook up to and adapt. The scientists themselves would plug in what they needed.' Chris Russell adds, 'They have the opportunity to hang things quickly and easily themselves from the cast-in Halfen channels in the concrete

EXPERIMENTAL RIG

A behavioural neuroscience experiment might involve a maze, an arena, or a box within which an animal makes choices; and one or more electrical means of recording their activity, including video. All this has to be connected to screens and computers to observe and record the results, while the level of lighting and sound may also need to be controlled. Because each experiment is unique, scientists usually build such 'rigs' themselves: some might fit on a bench top, others need a room-sized space.

soffit, without having to drill into the slab. Drilling into a concrete soffit causes noise and vibration throughout the building. So the idea is to allow the labs to change without disturbing the adjacent labs to the side, below or above.'

As well as the services supplied to the labs, and the equipment installed in them, the scientists will be able to change the way they use the space. 'We looked at using partition systems that are demountable,' says Russell, 'but in the end it probably takes as long to demount them and remount them. We have no ceiling because we want as much height as we can get for the scientists in between exposed services. Partition systems would need a ceiling. We also identified that plasterboard partitions were very economical, could be erected around the services when the labs are fitted out, and taken down fairly quickly. It's less expensive than buying a partition system that can be moved around. We visited labs that had partition systems, and they never moved them because it was too disruptive and too expensive, and they go out of manufacture after a while.'

As part of the planning for laboratory adaptability, the user representative Arifa Naeem worked with the lab consultant David Kelly to come up with a variety of possible lab layouts. 'I compiled a list of all the equipment that is used in neuroscience labs,' says Naeem. 'We sat down and went through all the different floors and we positioned the equipment. We came up with three different scenarios: one where all the labs would have benches, the middle scenario with half the labs with benches and half cellular lab rooms, and the third scenario was all cellular lab rooms. In the end it is up it is up to the director and the groups that come in to decide. That's the big feature of this new institute, the adaptability. Normally if you want to make an open space lab there's a lot more breaking down of walls, and the piping of gas and so on is not engineered for adaptability.'

Because the BSU and satellites have to be built to such strict criteria, the capacity for physical adaptation is rather less, but still significant. 'We've looked at when the building is actually in operation, what demands will be made on adaptability?', says Kelly. 'You can't start banging walls down when you've got animals in there, particularly with behavioural experiments. I was handed the basement to lay out as an animal facility with structural walls, and so you've got to work round that. But it's worked quite well. One issue in particular is how you segregate species such as ferrets and mice. Mice can sense they've got a predator next door. So we've put airlocks in. We might have zebrafish, so we can take another area up to 26°C.'

Similarly the engineering performance of the building was designed to optimise its potential. 'A lot of the performance criteria that we set were based partly on the known science, but also a lot about what might happen in the future', says Arup's Allison. 'So you set an extreme range.' There are limits, of course. 'You have to make assumptions about how people use the space' says DiMambro. 'There's a risk that you provide for every possibility, and then you end up with oversized plant. It's hugely inefficient – because you are running things at 20 per cent of what you have available. A lot of the assumptions that our team here make are based on experience and looking at what goes on in neuroscience: deciding how much data, how much power, compressed air, lab gases, everything. Making those assumptions is the key thing for me about adaptability. You find a middle spot, and allow some spaces to go to the higher end and some to the lower.' 'The demands of the structural and the MEP side always conflict', adds Allison. 'But the good news is that we are all one team here – we can balance the demands.'

Slowly, through a continuous process of investigation, discussion, decision, testing and refinement by the design team and scientists, emerged the design of a building that would work, within the parameters of performance and cost that had been set from the beginning. Now it was time to turn from the inside, and present the building's face to the world.

The layouts of the lab floors are highly adaptable

5

AN INTELLIGENT LOOK

THE SWC FACES THE WORLD

On 13 December 2010 a representative sample of the design and project teams, together with some of the scientific advisers, arrived at the Drill Hall in nearby Chenies Street (an arts centre since renamed as RADA Studios). On display for the public were drawings and visualisations of the proposed building, the first time neighbouring residents and others had had a chance to see what the new development might look like. Ian Ritchie and his colleagues steeled themselves for the kind of negative reaction that normally greets a modern building in a historic area. The massive, £660m UK Centre for Medical Research and Innovation (since renamed the Francis Crick Institute), under construction near St Pancras Station, had been given a very rough ride at its first presentation to the people of Camden.

But to their relief visitors, including representatives of local interest groups such as the Bloomsbury Conservation Areas Advisory Committee and the Fitzrovia Trust, gave mostly positive feedback. A month later Ritchie presented the proposals, still to be submitted to the planners, at Camden Council's regular Development Management Forum. Again the ideas were well received. 'I enjoyed that', says Ritchie. 'The UCL Estates team were a little nervous and thought it would be tricky, and we sailed through. We just explained our design and people understood it.'

While the eventual users of the building were most interested in all the imaginative solutions to the internal layout and adaptable servicing of the building that had emerged through the workshops, the building's neighbours were eager to learn how it would look, how it would fit into its environment, and how they

Concept drawing for the north façade

might interact with it. These aspects had all been formulated in the final stages of design development – apparently a contrast to the more conventional approach of designing the outside first and sorting out the inside later.

PUSHING THE ENVELOPE

The final design of the façade had emerged only towards the end of Stage D. 'I think we're in an age where a lot of architects think the skin is everything', says Ian Ritchie. 'The world of the image dominates us. One of the great characteristics of good architecture is that the inside belongs to the outside, and vice versa.' His insistence on getting the inside right first pushed the choice of cladding for the building later in the process than some of his colleagues were used to. 'We had a space plan', says Chris Russell, 'but we still didn't know how it met the outside world. At one point the scientists began to ask, "What is this going to look like?"'

The envelope, or façade, of a building has some practical jobs to do. Ritchie puts it very simply: 'It acts as a barrier to energy and water between the outside and the inside. Heat, light, sound, moisture, all those things that we sense – the effectiveness of the barrier determines both the comfort of those in the building, and to a certain extent the impact of the building on its external environment.' Yes, it gives a building much of its identity and gives the architect an opportunity to develop his aesthetic vision. But it can also make the difference between a building that is a delight to work in, and one that is unbearable. The design of the façade for the SWC emerged from an extraordinarily disparate set of thoughts about how the building would be used, how it would respond to its environment and how it might be perceived as part of the streetscape.

For Ritchie, although it was a modern building, it was essential that the SWC should respond to the heritage of the area. The surrounding neighbourhood is known as Fitzrovia and takes its name from Fitzroy Square, developed in the early 19th century as homes for wealthy families. The square's southern side is a splendid terrace of townhouses faced in white Portland stone. Ritchie began to wonder if he couldn't make another white building. 'Not the cream of Nash terraces', he says. 'Ice white is what I was going for.' He began to think in terms of an ice cube as a metaphor for the building. 'We were having to melt our own minds to understand what neuroscience was', he says. 'This melting we were going through to soften ourselves up, I thought was actually the same thing that the clients wanted

PUBLIC CONSULTATION
Applications for planning permission are public documents. Members of the public or other interested bodies can view the application and submit comments or objections to the planning authority. For major developments, it is common for developers to hold a public exhibition where neighbours and local interest groups can view the designs and ask questions before the detailed application is submitted.

Above Ian Ritchie's façade inspirations: water and ice
Overleaf Evening light on the completed façade

DISTOR
TION
ILLU
SION

to achieve between the different scientists – theoretical and behavioural – in the building.'

The material IRAL proposed to embody the metaphor of the ice cube could not have been further from Georgian stone or stucco. 'The thing that drove our thinking was the desire of the users to write on every surface', says Ritchie. 'So why not write on the inside of the building?' What was needed was a material that was strong enough to build a wall, yet writeable (and wipeable) on the inside, and the obvious choice was glass. 'Most buildings have glass', says Ritchie. 'It's quite cold, and usually has little texture. The irony of glass buildings is that from the outside they're black during the day, because they reflect light, and at night they are lit up because all the lights are on inside – and it is the lights that we see.' Glass had the advantage that it would allow natural light into the building. But could Ritchie avoid the aesthetic pitfalls of a conventional glass façade? Could it meet all the other needs for security and insulation? And was it affordable?

CLADDING
The cladding or outside wall of a concrete-framed building does not share the load of the building itself, but must control the passage of energy – heat, light, sound – between the building and the environment outside. It can be made of stone, brick, timber, glass, a variety of modern composite materials, or some combination of these. As the 'skin' of a building, it is a major factor in determining the structure's visual impact and environmental performance.

'Normally clients spend £1500 to £2000 per square metre on the façade', says Ritchie. 'I said "We're going to do it for £1000 per square metre." Bringing in that budget constraint would take us beyond the automatic, non-thought-through answer.' At the back of his mind was a project IRAL had delivered a decade previously, the Plymouth Theatre Royal's production centre, where they had wrapped large parts with woven phosphor bronze 'cloth', but also clad some areas with interlocking, U-shaped channels of cast glass filled with fibreglass insulation. It was strong, thermally insulating, and non-transparent, transmitting a diffuse light that did not cast strong shadows to the large paint studio. And it looked white from the outside. 'We thought we could get the thermal and light qualities; we knew we could make it watertight', says Ritchie. 'Now we were going to make it much bigger.'

The cast glass channels are made by a family firm, Glasfabrik Lamberts in Bavaria in southern Germany, who had supplied the cast glass for the Plymouth building. Ritchie began to collaborate with both Lamberts and UK-based Pilkington, to achieve the desired result. Each built sample assemblies, and samples of the glass channels were brought back to the studio to test. Finally, the architects chose the preferred assembly of low-iron glass and optically white insulation which gave a truly white finish. 'We looked at all the standard textures, and the

one for the outside that appealed to me had a finely ridged, wave-like profile,' he says. 'It struck me that it would conduct the rainwater and that made me think, "We won't have to clean the glass." Just like stone, you'll do it once every ten or twenty years.'

The Howland Street façade is north-facing. A glazed finish to the south side could make the building uncomfortably hot during the day. 'Historically people weren't bothered', says Ritchie. 'You put a blind in and you still cooked.' But he realised that the laboratories that occupied much of the south side of the building did not need light: working with microscopes or nocturnal animals, scientists need a dark, enclosed space. So the south façade could be a conventional insulated wall. That made it possible to save space in the building by taking ventilation ducts and other services outside. However, even though the south side does not directly face a street, Ritchie was not satisfied to leave a wall criss-crossed with ductwork, and began to think about how it might be screened to provide an aesthetically pleasing urban backdrop.

'We decided to work with nature, and put in something that moves', he says. Going back to the concept of the building as an ice cube, he proposed a fractured 'veil' of suspended, thin, rectangular fragments or 'pixels' – a metaphor for a field of ice floes – that would provide shade and ripple gently in the wind. This 'lamellar screen' gave him another idea about the north side. 'If you're going to have movement on one side, is there a symbolic interpretation to do with the wavelengths of signals

The lamellar screen ripples in the wind

in the brain?' He began to see the Howland Street façade not as a flat surface, but as a series of vertical waves.

'That rhythm applies to how you look at it in the street', he says. 'You will never look at the main façade of this building perpendicularly, always tangentially. Because you're in a Georgian area of London, surrounded by a conservation area, getting verticality into the design is important. All visually horizontal buildings within the city look wrong.' The rhythms introduced by constructing the façade as a series of waves mimic the verticality of traditional Georgian buildings. 'The Georgian rhythm – the spacing between party walls in the Georgian terraces of London – is roughly 6 m', says Ritchie. 'Even though traditionally brickwork is a single plane, the windows give you the verticality, and the door indicates the separation.' Partly as a joke about the rapidly firing neurons of the computational neuroscientists, and partly to give the GCNU a distinct identity from outside the building, he increased the frequency of the waves across the central block of the building. 'There are always differences [in Georgian architecture] that make the places interesting', he says. 'I thought that while using only one material, we would vary the frequency, and enjoy the material at the small scale as it mimicked the larger waves which it was creating.'

The walls at the eastern and western ends continue the whiteness and translucence of the cast glass, though in a flat plane. At each end of the laboratory levels, where there is a meeting room, there was an opportunity to provide views up and down the street, and so they are faced in clear glass. 'It looks as though when the cast glass, which is ice, comes round the corner, it melts into liquid' says Ritchie.

Much of the detailed design and research on the building envelope was undertaken by one of the younger members of IRAL, Karl Singporewala. The architects worked with a specialist façade contractor, Frener & Reifer who are based in northern Italy, to develop the finished design. 'It's been very interesting working with the Italian façade contractor', says Singporewala. 'They have brought a passion and rigour to the detailed design, working closely with us. They were selected because of their understanding of our design. They produced their own tools and said, "OK, this is how I think you would make that." And some things we agreed with and some things we didn't, and we learned from each other.'

When they were first presented with the design of the façade, the scientific advisers, inevitably, were less concerned with its appearance than with whether it would perform as required. 'The deal was that Ian could have the façade', says

Rear elevation: in the foreground, a glass skylight allows light into the basement staff room

John O'Keefe. 'It satisfies a lot of conditions, it is adventurous, but it's sympathetic. It has a better level of insulation than most buildings. And they've acceded to my demands for windows that open.'

The arrangement of the windows has provided another opportunity for a changing rhythm across the north elevation, again responding to the needs of those inside. Windows are more densely clustered across the central section occupied by the GCNU than they are in the laboratories, and placed at more frequent intervals in the upper than the lower lab levels. 'There's an extra layer to the façade that we won't see until people are in the building', says Singporewala, 'which is the motorised louvres to all the windows.' Each window and louvre has its own control: it can be open or closed at will by the occupants. 'People who have their own offices will be more sensitive to being able to bounce evening light off the louvres into their office spaces,' says Singporewala, 'compared to the open write-up spaces where a lot of people might be clustered around one window. 'The play of the louvres may be reminiscent of the keys of a piano being played', Ritchie adds.

The innovative use of structural cast glass and moving panels has created a building that looks unlike any other: yet, according to the Ritchie, it has been derived from London's and Fitzrovia's built history. The clients were not looking for an iconic building, but they may well have got one anyway.

IRAL was able to fulfill the scientists' desire to have windows that open

ENGAGING THE PUBLIC

A research laboratory is not a public building, and traditionally labs made very little attempt, through their design, to communicate with passers-by or invite them to enjoy the building. Ian Ritchie was keen from the start that the building should engage with the wider community. 'Howland Street is actually an outdoor room', he says. 'The difference here is that the wall that was outside has now become inside. The street is a public room.'

The SWC's predecessor, the Windeyer Building, had stepped back from the pavement, and people passing by were separated from it by a wall and a fenced area of parked cars. The entrance was tucked away around the corner on Cleveland Street. The SWC was designed with a larger footprint, taking it closer to the street edge. 'How you entered the building, and how the building met the ground, evolved', says Chris Russell. In an early design the entrance was at the eastern end, with a section of the ground floor stepped back to form an undercroft. But by the time the design was ready to go to the planners, internal uses had been reorganised to allow that idea to be extended into a full colonnade along the front of the building. 'The generosity of the pavement is something I've always believed in', says Ritchie. 'It goes back to the idea of the outdoor room. The space you make for the threshold of a building, not just its entrance, says you're either welcome here or you're not.' It was an intentional urban move – a generous gift to the passer-by.

The colonnade is a public space, providing shelter from the elements in the manner of the decorated arcades of Italian renaissance cities such as Bologna. There was some anxiety during a public meeting that the sheltered space might prove a magnet for rough sleepers, but the planners were eventually convinced that this would not be an issue.

Further developments in the ground floor façade have provided more invitations to passers-by to interact with the building. The eastern half, which includes the lecture theatre and the main entrance, is glazed in clear glass, as is the southern side of the reception and a portion of the lecture theatre area. Anyone passing can see right through the building, and where appropriate can see what's happening in the lecture theatre (there are blinds that can be drawn to control daylight or reduce distraction). The entrance is via a revolving door, kept unlocked during daylight hours: further secure entrances next to the reception desk control

POCKET PARKS

City authorities are conscious of the need to preserve green space in urban neighbourhoods. Part of the Mayor of London's 'Great Outdoors' initiative, pocket parks are small areas of 'inviting public space' that improve the experience of people walking through London's streets, squares, parks and waterside spaces. They must include trees and other planting, and places to sit and reflect. The aim was to create 100 such spaces across London by the spring of 2015, making the city 'friendlier, greener and more resilient.'

Transparent glass on the corners gives the impression of ice melting into water: pocket park at street level

access to the laboratories. This arrangement makes it possible to use the lecture theatre for public events. At the same time, after discussions with the planning authority about providing some public amenity space, the whole western end of the building has been set back 10 m from the edge of Cleveland Street, allowing space for a public 'pocket park' and complementing the one diagonally across the junction.

ART FOR SCIENCE'S SAKE

The budget for the SWC provided for an unspecified 'art project' to enhance the building, and developments in the design of the building envelope crystallised Ian Ritchie's ideas for how to fulfil this part of the brief. The architects made an early decision to puncture the opaque wall facing the colonnade with five full-height clear vitrines, which could display a changing array of exhibits. 'There's the chance to have a communication of the building's activity sensed and read by the public, which very few buildings achieve', says Ritchie. 'Interestingly all shops sell you what they have through the window. But not many science buildings communicate what they are doing.'

The content of the various elements of the art project was not finalised until long after the planning application had been accepted – indeed, until the building was almost complete – though Camden stipulated that each part of the project must subsequently be submitted and signed off. It was David Sainsbury's strong preference that the contents of the vitrines should be based on scientific principles, rather than Ritchie's original proposal of contemporary art pieces inspired by neuroscience. Thinking back to the stimulating sessions he had enjoyed with Richard Gregory at Cambridge, Sainsbury suggested that they should focus on perception and visual illusions, which are captivating for the viewer but also tell you something interesting about how the visual system works. And through Sarah Caddick he also provided an adviser for the project.

Marty Banks is a professor of optometry and vision science at the University of California at Berkeley. 'Our job is to make people understand how difficult it is to see and interpret things', says Banks. 'Illusions are a very useful way of doing that.' Before the SWC was launched, Banks chaired a committee reviewing the work of the GCNU. 'I was in London for two or three days, and met Lord Sainsbury at the end of it', he says. 'I had a chance to chat with him, and he ended up coming to Berkeley and touring the lab. We said, "Maybe we'll get together on some project", and so it has turned out.'

On the same visit Banks introduced David Sainsbury to a colleague, Hany Farid, a computer scientist at Dartmouth College in New Hampshire. 'Hany's specific expertise is in image manipulation', says Banks. The two scientists worked together on types of illusion that might work in the SWC's vitrines, such as simultaneous contrast and shape from shading; as the building neared completion they added a third member to the team, the exhibition designer Maria Mortati from San Francisco. The exhibits have to be appealing but also informative to curious passers-by. 'There's a moment when you have their interest and attention,' says Banks, 'and you want to take advantage of that, explain how the process works, and hopefully engage people for the future.'

The colonnade soffit also provided an opportunity for Ritchie to show his relaxed and poetic side, again assisted by Karl Singporewala. 'The decoration of the soffits was about going back to the renaissance in Italy', says Singporewala. 'You don't see that in London.' They soon rejected the initial idea of installing a ceiling of laminated glass panels encasing imagery below the soffit (too likely to cause collisions between distracted walkers). Instead, the IRAL team realised that the soffit was high enough to suspend more of the composite 'pixels', used for the lamellar screen on the south side of the building. They would provide the opportunity to create mosaic images, visible without craning your neck, that would change with the perspective of the viewer, and move with the wind to reveal an aspect of the changing environment.

A colonnade invites passers-by to enjoy the building

J S BACH 'MUSICAL OFFERING' (1747): RICERCAR A 3

'The distillation of three perfectly interweaving lines, of shared melody in perfect form and harmony, reflects the ultimate in unveiling the mysteries of the musical mind. Bach's famous visit to Potsdam to visit Frederick the Great on 7 May 1747 inspired a theme in his 'Musical Offering' which conflates the Renaissance knotty-ness of the old Ricercar with the staged luminosity of the fashionable Enlightenment. As artists so often do, the return to first principles at the end of a creative life leads to works where not a note, a word or a figure could be removed without reigning incoherence. Bach's raw scientific data absorbed and lost into the realms of the ear, heart and mind is just the start of a 40-minute work which hurls the 16th-century towards the late-19th in a single arc of creative genius.'

Jonathan Freeman-Attwood, *Principal, Royal College of Music*

'Pixels' suspended from the colonnade soffit illustrate the brain's creativity with a Bach score

On the reverse of the 'pixels' are composite portraits of UCL's Nobel prizewinners: intriguing visual illusions punctuate the façade

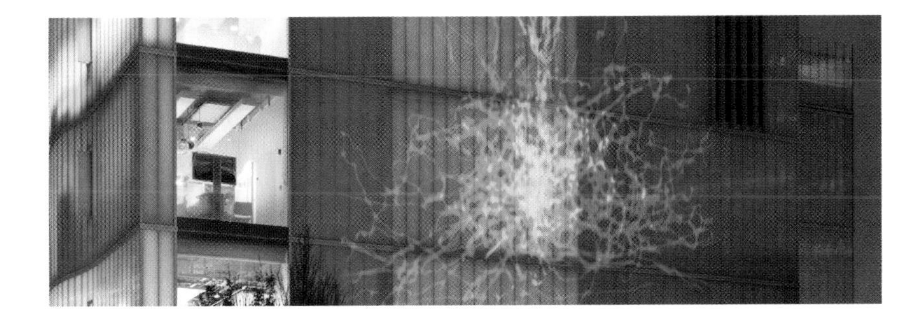

The 1000 pixels are suspended vertically at regular intervals perpendicular to the façade. They are printed with images that evoke the creativity of the mind. Passers-by see the score of J S Bach's 'Musical Offering' – occasionally called the 'Prussian Fugue' – when travelling from east to west. As they travel in the other direction, they see composite portrait images of Nobel prizewinners. 'The Nobel prizewinners are all in Physiology or Medicine, and all affiliated with UCL', says Singporewala. 'Originally there were eleven of them, and we had twelve colonnade bays to work with, so there was one bay that did not have a portrait in it.' The idea was to challenge the occupants of the building with the question: 'Who is going to be the first scientist from the SWC to have that space?' The answer was to come sooner than anyone expected (see p. 158).

The five-storey, white, western end of the building also presented a blank canvas that invited some artistic response. When Ritchie met resistance from the key scientists to the idea of a fixed representation, he proposed instead to project a massively magnified piece of imagery from neuroscience, such as a neuron lit up with fluorescent labelling. The flat roof of the delivery entrance on Cleveland Street proved ideal as a place to mount the projector, while an etched information panel in the pocket park could explain the image, projected on the wall at a size of 10 m x 10 m, to curious passers-by. The beauty of a projection, of course, is that it can be changed in response to the changing research focus of groups in the building (though Camden would insist on vetting each one).

The final element of the art project was the book that you now have in your hand. Ritchie, a writer, was keen to document the process, rather than celebrate the architecture. He wanted others to understand how new ideas emerge, that innovation is not something to fear, and how important different people become at different stages of realising a building.

The west face of the building provides a screen for projected neuroscience images

ACCORDING TO PLAN

The client team submitted the planning application to the London Borough of Camden on 25 February 2011, with all the design features apart from the detail of the art projects decided. An initial stumbling block was Camden Council's policy that such new developments had to include affordable residential housing proportional to the size of the building. UCL's Provost, Malcolm Grant, was adamant that this was not appropriate. 'It's not sensible to tax one public good to provide another public good', he says. 'We were determined not to do it on site – it would have set a terrible precedent.' Eventually Camden accepted a compromise: UCL agreed to add two extra floors to its existing student housing scheme at King's Cross, which was linked to the SWC project through a Section 106 agreement. The pocket park and public art were similarly included as obligations, but as they were part of the design anyway this did not cause any difficulty.

The application came before the planning committee on 21 July 2011. 'We all went along that evening', says project sponsor Stuart Johnson. 'It wasn't widely attended, but neither was it a closed meeting. Ian was asked one or two questions, but the officer gave it a recommendation. Three councillors rejected it or abstained, and everyone else supported it.' The next day UCL issued a press release announcing that the project was going ahead. John O'Keefe, then Interim Director, added the quote: 'Everyone involved in this ambitious project is very pleased to have passed this important milestone. In addition to providing a world-class neuroscience research centre, we believe that the building will add significantly to the aesthetic quality of the area. I want to thank everyone involved for their hard work in getting this visionary project to this point.'

The successful application was indeed a major achievement, but so carefully had the team conducted its local consultation that news that the building would go ahead attracted little public comment.

PLANNING PERMISSION
Any design for a new building or a substantial alteration to an existing building has to gain the approval of the local planning authority, usually the local council. The authority's development plan will place constraints on the types of development permissible (business, residential, educational and so on). It will also provide guidance regarding the height and density of buildings, and will apply standards on environmental and energy conservation grounds.

FROM IDEA TO REALITY

By the time the planning application was submitted, a main contractor to carry out the building works was already lined up. The team from Kier Construction, who had recently completed the Sainsbury Laboratory in Cambridge, was among those who tendered for the job towards the end of 2010. Kier was selected as preferred bidder in March 2011. 'I had previously worked on two hospitals,' says Nick Mann of Kier, who was the project manager on the SWC, 'which are also highly mechanically and electrically serviced buildings. The attraction for me of the SWC was the added dimension of architecture. Hospitals don't have the same level of architectural vision.'

From this point until the construction contract was granted Kier would work with the design team to prepare the tenders for the trade contractors, and draw up plans for the demolition of the Windeyer Building. 'From March to December 2011 we worked with the design team to understand the design', says Mann. 'We then went to the market to check it was affordable. The contractor has input to the design's buildability to make sure it works.'

In June 2011 the last occupants left the Windeyer Building. It had stood for 52 years and witnessed remarkable advances in studies of immunity and of viruses such as HIV and hepatitis B. Its architect Teddy Cusdin had used the same phrase as Ian Ritchie to describe his approach – designing 'from the inside out' – and its long spans with no internal supporting walls, offering the potential for adaptable floor arrangements, had been innovative for the time it was constructed. But it had begun to show its age, and a question mark had been raised over its future even before UCL entered its bid for the SWC.

The hoardings went up around the building by the end of June, while the design and construction teams waited for the planning decision. Once the green light came in July, the task of demolition began. 'Taking a seven- or eight-storey building down in Central London in a highly populated area is challenging', says Nick Mann with heavy understatement. 'It turned out to have a lot of asbestos, which took time to remove. There were extensive foundations: 11-12 metres down. There were financial consequences, but it fitted into the contingencies allowed.'

It eventually took almost a year to clear the site. Meanwhile Kier was working with IRAL to cost the designs and recruit a team of specialist contractors. In May 2012 Kier signed the main construction contract with the funders to build the SWC at a cost of over £70m. At this point IRAL and Arup were themselves novated to Kier as architectural and engineering consultants.

With an eye to achieving a high level of energy efficiency and sustainability, Kier began by crushing some of the old concrete from the demolished building to reuse for piling mats – areas of hard standing for the heavy piling rigs. These now began to drive in the interlocking secant piles that formed the retaining walls below ground, and the load-bearing piles that would support the vertical

MAIN CONTRACTOR
The main contractor is the building firm appointed to turn the design into a finished building. A 'design and build' construction contract is usually signed as soon as planning permission has been granted, but the contractor may have been selected as 'preferred bidder' in advance, and will work with the architect on the 'buildability' of the design and on preparing the tender documents for trade contractors. This enables the main contractor to establish a price prior to the transfer of risk from client to the main contractor.

columns of the building's frame. With the site's boundary secured, the contractors began to excavate to the depth needed to accommodate the two basement levels in the design. 'Digging down 12–13 m to create two floors below ground in Central London was both time-consuming and expensive', says Ian Ritchie, ' but the only way to achieve the floor area in the brief was by going down.' A further challenge was the proximity of the substructures of neighbouring buildings – the BT building immediately to the north (which incorporates the iconic 1960s Post Office Tower, London's tallest building until 1980), is on a very large raft foundation.

Big construction projects can be noisy and disruptive, and dealing with the reactions of the neighbours was one of the UCL Estates Office's concerns. Immediately adjacent to the site is Astor College, a hall of residence for UCL students. The students were justifiably concerned about the dust and disturbance that would begin with the demolition and continue throughout the construction period. 'Kier held monthly meetings with the students and provided a number of bottles of beer', says Dave Smith. 'We didn't get a lot of complaints, and the rent was reduced to compensate them for the disturbance.' Some of the neighbourhood associations were also concerned about noise and construction traffic. But in the event Kier's concern to be good neighbours paid off. They circulated regular newsletters, and held monthly open meetings when anyone could come along and ask questions. As far as Nick Mann recalls, only two people ever turned up in two years.

DESIGN AND BUILD
Design and build is one of a number of procurement routes for buildings. Under such a contract the main contractor has overall responsibility for the design and construction of a building, working to the employer's requirements and an agreed maximum price. The contractor takes on the financial risk of design development or other delays to construction. On signature of the contract the architect may be 'novated' to work as a sub-contractor.

To take just a single example of the work of the trade contractors, the Italian firm Frener & Reifer were appointed to construct the building's façades to IRAL's innovative designs. Ian Ritchie has worked with many of the technically advanced façade companies in the world, particularly in Europe, and is renowned for his structural glass designs. But the cast glass panels of the north façade, with its wavy profile, had never been used before in quite this way. Frener & Reifer were equal to the challenge. 'The contact with the trade contractors is the bread and butter of getting what we want translated into reality', says Gordon Talbot of IRAL. 'Frener & Reifer understand what they are doing, they don't deal with uncertainty.' In association with IRAL, the contractor carried out extensive tests on the panels for their resistance to noise, moisture and impacts long before they were installed. And the installation process was seamless. 'The panels all arrive as a kit on the lorries,' says Talbot, 'and when you unpack it, it's

all in the right boxes in the right layers, meticulously labelled. That understanding of rigour and logistics is something that is less well understood or commonplace in the UK.'

The façade was installed during the latter part of 2013, and Karl Singporewala, who had led the façade project for IRAL, immediately began to appreciate the way it responded to its environment. 'Since it started getting darker I've been coming to look at it while the construction lights are still on,' he said at the time, 'and though I say it myself it is quite beautiful. The soft light is coming through, but interestingly the lighting on the street outside and the cars going past are throwing red and orange light all the way up to the fourth floor, which is welcome. I've had people say "Is it plastic, is it stone?" You only get that real understanding from it at night time.'

TRADE CONTRACTORS
On a large building project the work may be carried out by a number of contractors who specialise in particular trades: groundworks, concrete structures, steel fabrication, brickwork, services, façades and so on. Under a 'design and build' contract, they are all subcontracted to the main contractor rather than the client.

FIXTURES AND FITTINGS

To help the scientific advisers understand and appreciate the IRAL design and products, in the autumn of 2011 a 'lab prototype' was constructed in a warehouse in Croydon. Representing one structural

Installation of the prefabricated cast glass façade units

After dark the translucent façade complements the lit streetscape

bay of a laboratory floor, the prototype was serviced and finished as envisaged in the building itself. It was then subdivided into two cellular lab rooms and furnished with plywood representations of lab furniture and equipment. It played a valuable role in developing the final stages of the interior design and servicing. For example, the scientists had been particularly concerned that they would be able to disable automatic lighting controls, such as daylight sensing and presence detection, to avoid disruptive changes in light levels during experiments. During visits to the prototype they were able to reassure themselves that the lighting system was suitably simple, controllable and adaptable, while the design team were able to test the lighting for electromagnetic and ultrasonic emissions. The prototype also gave the services contractors the opportunity to try out the installation of the prefabricated, exposed service runs.

Prototypes provided an opportunity to demonstrate the fixtures, fittings and furniture that would make the building into a functioning working environment. Decisions about the look and materials of every item in the building were all regarded as part and parcel of the architects' work. 'My view, supported by Susie Sainsbury and by Gatsby, is that you don't design by committee', says Stuart Johnson. 'The finishes and the furniture and the art are all facets of the same thing – it's architecture. You've chosen a world-class architect, and unless they've proposed something that's unbuildable or unusable, you just accept their advice.'

Much of the office furniture was bespoke, designed by Ian Ritchie with Teresa McQueen of IRAL in collaboration with the specialist contractor Luke Hughes & Co Ltd. Luke Hughes specialises in the design and fabrication of furniture to enhance both ancient and modern architectural spaces: his long list of clients includes Oxford colleges, cathedrals, hotels and corporate offices. 'The main reasons that we are custom making the furniture are to fit within the visual context, to meet a budget and to get good quality', says McQueen. 'We come up with a general design idea and we work with Luke Hughes on details, materials, costing and so on.'

The colour scheme was predetermined. Ian Ritchie stuck with his ice-cube metaphor, stipulating that just like the exterior, all the interior walls should be white. He did, however, want to introduce a second colour, and felt that the only one that went with the

A lab prototype made it possible for the team to review the colours, finishes and service installation before construction began

ice cube concept was blue. 'A particular blue happens to be my favourite colour', he says. In one of those serendipitous encounters that emerged from the long research period at the beginning of the design process, he had met the Professor of Circadian Neuroscience at Oxford University, Russell Foster, who was a member of the SWC Governing Council. 'I met him for the first time at the Royal Society,' says Ritchie, 'and after about half an hour we went for a coffee upstairs and he asked me "What's your favourite colour?" And I said "481 nanometres". And he said, "First of all, you're the first non-scientist that's ever responded with wavelength, and secondly you're only one nanometre out from that which makes you most alert.'

On the spectrum of visible light, 480 nm can be found in the intense blue of the sky. Ritchie then found a blue paint for concrete surfaces in a hue that looked about right to him, so he asked Russell Foster and his colleagues to come and test it. 'He came up with a spectrometer, and it was 480 nm. So I asked him "Do you think if we put this as a ceiling in the labs, broken by the silver grey acoustic panels partially obscuring the services, will that be a relaxing background and good for scientists?" And he said "Yes." And we took the decision to paint all the soffits blue.'

There was some wariness in the workshops, but John O'Keefe thought it was an interesting idea, and so did Richard Morris. The contractors painted the soffits in the lab services prototype blue, and no one objected. Ritchie has no doubt about the result. 'I think it's worked brilliantly', he says. 'You can sense that we've done a bit of homework. It went with the ice, and went with the science.' The vinyl flooring is also a deeper tone of the blue.

The choice for the furniture, against the dominant blues of floor and ceiling, was more white. 'Visually, it's not noisy', says Ritchie. 'This building should be calm – it should have a silence. People bring colour. They're the dynamic things, because they are moving all the time. It's very rich.'

For the more junior researchers (postdocs), guided by Ian Ritchie, McQueen and Hughes designed groups of individual pods or carrels – all in white of course – with curved desks and shelving, that wrap around to provide a sense of privacy even in the open plan space. 'The space is quite tight', says McQueen. 'To get 30 people in, you need to have a compact arrangement. There are also acoustic concerns, so we've provided a desk that is quite tall.' The backboard is finished with a magnetic 'pinboard' that conceals a layer of acoustic absorption. 'Originally we had quite a rectilinear approach' says McQueen. 'Luke Hughes suggested a more curvilinear approach which had advantages in terms of manufacturing techniques and cost – and also reflected the architecture, picking up on the theme

Above and right Purpose-designed modular 'pods' or workstations for graduate students and junior researchers

of the curving north façade.' The curvy front edge also optimises the useful desk space, while the 120-degree angle at the back of each carrel means that they can be arranged in compact groups with varying numbers of elements. The assembled carrels create individual space without walls.

SWC's interim Centre Manager, Marg Glover, was a key member of the team scrutinising the practicability of the designs. 'Each person has their own little space', she says. 'It's quite tight, but they've demonstrated that they can put two big screens and a laptop on each one. John was very nervous about open plan being too noisy, and not being able to focus. But we went round and talked to people in open plan offices at UCL, and in each case they said they enjoyed the open plan atmosphere. John was pleasantly surprised!' To avoid disturbance, and because nearly everyone uses email or mobile phones to communicate, landline phone extensions will not be installed in the carrels.

'A question that came up was "Where are they going to hang their coats in the write-up area?"', says Glover. 'The design team said "Oh, they hang them downstairs, in their lockers." They're half lockers – that's not the answer. You will end up with coats and other items strewn around and on the backs of chairs.' There are now integrated coat-hooks at one end of each of the carrel runs.

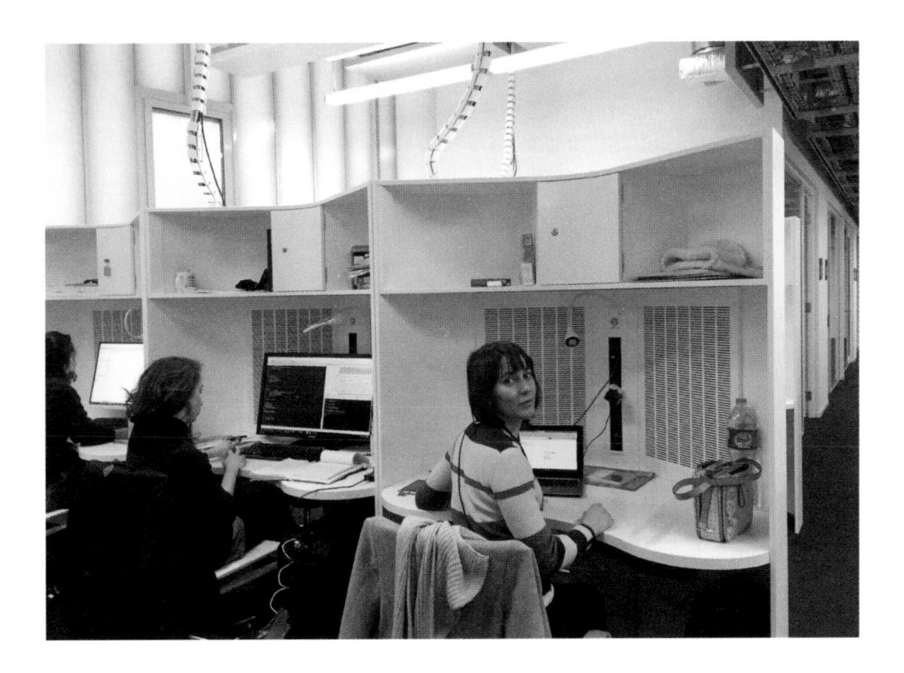

Getting the furniture and fittings right is not just a matter of pleasing the users or creating a consistent aesthetic. Wherever animals are involved, stringent requirements for hygiene and safety have to be met. When Glover went with IRAL to review a prototype of the changing area joinery for the staff of the BSU, she immediately asked for changes. 'Since the original discussions of what was needed, three years before, things had changed dramatically in terms of the PPE [personal protective equipment]. It used to be that everyone hung a lab coat up, and put it on when they went into the laboratory. Now full coverall is a requirement. So we had lots of hangers to hang all the lab coats, and I said, "That's not what's going to be required when we go in the building."' Glover was able to specify exactly the size and shape of storage that would be needed for supplies of coveralls. 'And that's really very useful,' she says, 'because we would have ended up with a unit that wouldn't have worked if we hadn't had that discussion.'

This attention to detail extends throughout the lab furniture for the BSU. 'We've got a furniture design that has the best of all worlds – it looks smart and it's functional', says Glover. 'Units that are going into the BSU have to be designed so that the animals can't escape and get behind the units. Mice can get into the most amazing places. The Home Office is very strict – their level of cleanliness far outweighs what you'd see in your kitchen. High-level surfaces have to be sloping so that dust doesn't accumulate, there must be no cracks, no sharp edges, the join between wall and floor has to be very smooth, and the benches have to fit against the walls perfectly so that they can be sealed all round. The surfacing of the units must be chemical-resistant and easy to clean. The cupboards and drawers are all on casters so that they can be rolled out to clean behind.'

Glover herself appreciates the attractiveness of the building, but at the end of the day its success will depend on these functional details. 'The thing that's going to be important is that everything works. If you walk in and the services don't work, or you can't get the temperature right, this will have a major impact on the functionality of the building. It's nice to have a pleasant environment – everyone's excited to go into a newly-built lab. But what's also important is that you can bring a visitor in without any hassle, you don't have to jump through hoops for everything you want to do. We have to make sure that research is supported in a seamless way. Yes, it's more about human interaction than the building, but if the building's badly designed it's not going to work.'

On 26 June 2013 Gatsby, Wellcome and UCL held a 'topping-out' ceremony to mark the completion of the building's structure. In attendance were Lord and Lady Sainsbury representing the Gatsby Charitable Foundation; The Wellcome Trust's Acting Director Ted Bianco; UCL's Provost, Sir Malcolm Grant; Colin Lamb, Managing Director of Kier Major Projects; and John O'Keefe, who just the week before had been confirmed as the Inaugural Director of the Sainsbury Wellcome Centre. The band played an extract from Bach's 'A musical offering', the score of which Ian Ritchie had chosen to depict on the 'pixels' decorating the building's pavement colonnade. Everyone expected that a year later, the building would be complete and the scientists would be moving in, ready to put the architects' innovative designs to the test.

Leaders of the client and construction teams at the 'topping out' ceremony on 26 June 2013

6

MOVING IN

THE LABORATORY COMES TO LIFE

On 23 May 2016 the SWC opened its doors to a select guest list, bringing the great and good of the British scientific community together with those who had brought the building into being. Eric Kandel had flown in from New York to give a keynote lecture that would mark the Centre's formal opening. It was an opportunity for many of the characters who fill the pages of this book to enjoy the building in its finished form. And for guests it was a first chance to glimpse the unique interior and appreciate a building created entirely with the brain in mind.

The lecture theatre was filled to capacity, with students lining the back wall to hear Kandel's presentation. There was no one better, as John O'Keefe said in his introduction, to open the Centre: Kandel has had a lifelong interest in neuronal circuits, and made seminal discoveries about the molecular basis of learning and memory that brought him the Nobel Prize for physiology in 2000.

Calling on his audience to 'imagine the rejoicing of the angels at the opening of this building', Kandel paid tribute to the 'extraordinary tradition of British brain science', which he traced from Charles Sherrington's work on reflexes in the 1890s directly to John O'Keefe's discovery of place cells in the 1970s. Moving to research on learning and memory in his own lab, he demonstrated that long-term memories depend on the active synthesis of new proteins. 'If you remember anything about this lecture', he wisecracked, 'you will walk out with a different brain than you came in with.'

Eric Kandel celebrates the opening of the SWC

UCL President and Provost Michael Arthur thanked Kandel for 'setting a wonderful tone for the future of the SWC', while Wellcome Trust Chief Executive Jeremy Farrar praised the 'vision, fantastic working environment and great leadership' that had come from David and Susie Sainsbury's commitment to the project. It would enable young scientists, he said, to 'dream beyond where academic research normally goes.'

For John O'Keefe, it was an opportunity to reflect on what had been achieved in the eight years since he and his UCL colleagues first dared to dream. In 2014 he had received a Nobel Prize for his work, and with a full programme of research still to undertake he was ready to hand over the administrative burden and move back to full-time research. From the end of September 2016, someone else (yet to be announced at the time of writing) would be taking over what he had steered to completion.

David Sainsbury, who is thrilled with the building, is confident that it helped to attract a strong field of candidates for senior positions. 'If you're going to recruit in a market where there's a lot of competition', he says, 'having outstanding facilities in an attractive building is very useful. Working in good conditions makes people more productive, and how people interact is very important. We've put the computational neuroscience people in the centre of the building, because you'll get the interaction between the experimentalists and the computational people. It will work both ways: you will have computational people saying, "Look, theoretically this is how you can do this, but do you have any experimental evidence?" And equally you will have the experimentalists saying, "We've got these results, what does that tell us?" The interaction between the two is fundamental to cracking particular problems.'

The building had been due for completion in 2014, a year from the topping-out ceremony in the summer of 2013. Unfortunately Balfour Beatty, the services contractor, had considerable problems accessing resources for their industry, leading to a significant delay. For project sponsor Stuart Johnson the delay was a serious problem, and one that had to be managed pragmatically. 'Because of the way I structured the contract', he says, 'none of this cost the funders anything. I'm personally pleased that it was not a disastrous delay.'

Kier finally handed over control of the building on 24 November 2015. By that time John O'Keefe was ready to take up residence in the Director's 'room with a view' on the fifth floor: key members of the Centre's administrative team moved in at the same time. 'The early word on the street at the start of the project was

Previous spread Reception area with adjacent lecture theatre and outside terrace

that it was going to be an iconic building expressing the architect's ego,' says O'Keefe, 'but it turned out not to be that at all. They have designed and built a building that will be first-rate for science. Ian is an architect who believes that you build buildings for a purpose. This whole functionalist view, which I thought was obvious, has kind of died in architecture practice. So we are very lucky. If the building fails it will be my fault for not having foreseen the future challenges!'

Centre Manager Alexandra Boss is particularly delighted, having spent the previous 18 months in a small room at the back of O'Keefe's old lab in the UCL Anatomy building. 'There's a massive feeling of space', she says. 'It's an extraordinary privilege to have a building like this. In an academic environment,' she adds with heavy understatement, 'the accommodation is often not ideal.' With her colleague Mark Dixon, Operations Manager for the SWC, she had gone ahead with letting the security and cleaning contracts before the building's official handover. Taking partial use of the building in this way – which is unusual, but allowed if contractors run late – gave time for complex systems such as the security doors on the ground floor and the air handling systems to bed down, and to show up any problems that needed sorting out. 'Over the past two years,' says Dixon, 'the most significant problem has been the commissioning of the various systems and the complexity of how they communicate with one another. A positive from the delay in the handover of the building has been that it has allowed me and my team to get to know the systems well.' Meanwhile a team from the Home Office carried out a final inspection of the impressively finished BSU and its lab satellites and confirmed that the spaces could be licensed.

Peter Dayan and the GCNU – a team of 25–30 people – had moved the mile or so westwards from Queen Square and occupied the central labs of the new building at the end of July 2015. The lecture theatre began to be used to hold talks during the autumn term. Yes, there were still contractors on site, but as young researchers cycled in, and the GCNU kettle was getting up steam, the building was beginning to feel more like a working institute.

Dayan expresses some regret at leaving Queen Square, and the 'cosiness' of the building they formerly occupied. But on the other hand, he says, 'Many things here just work so much better!' The GCNU combines a need for monk-like seclusion with a busy programme of events when they get together to share ideas. The graduate students have settled into their four-person offices, and immediately

WHAT IS THE NEURAL CODE?
Everything we experience in the world, and everything we do in response, is encoded in the brain as electrical pulses. These pulses flow through neural circuits, conveying sensory information, communicating between brain areas, and sending commands to our muscles for movements. How the pattern of pulses encodes information is still unknown. By simultaneously recording from many neurons in many brain regions, SWC scientists will try to crack the neural code.

Façade interior serves as a writeable surface, immediately adopted by the occupants

adopted glass surfaces as whiteboards, as intended. At the centre of their new space is the all-important seminar room, with the even more essential tea making area. Until the rest of the building was occupied it would be difficult to tell if the opportunity for SWC members to 'eavesdrop' on seminars from the upper level would be taken up: however, the glass windows into the space remove any sense that the GCNU is hidden away.

Dayan is particularly pleased with the provisions that have been made for computing. 'Before, we had a broom cupboard stuffed with all kinds of air conditioning and a prayer that the air conditioning didn't go wrong', he says. 'Now we have a proper, air-conditioned machine room. So we can expand our computer facilities.' John Pelan, who was formerly the senior systems consultant to the GCNU, has been appointed head of scientific computing for the SWC as a whole: it is largely down to his planning that the transition has worked so smoothly. He will run an integrated service for the GCNU and SWC, bringing greater efficiency and economies of scale.

More formal events have gone well, says Dayan. 'When we have seminars in the lecture theatre on the ground floor, it's been lovely to have our post-seminar refreshments on the terrace at the back, especially in the summer when we first moved in.' The GCNU brought in caterers to host a big event in the autumn of 2015: the quinquennial review of its performance that feeds into the next round of funding from Gatsby. 'We held a 100-person workshop and then a number of dinners in the brasserie upstairs', says Dayan. The success of the event bodes well for the annual meeting Gatsby organises between the GCNU and theoretical neuroscience groups that it funds at the Hebrew University of Jerusalem and Columbia University in New York, scheduled to be held in London in 2016. Both are also due to move into new brain sciences buildings, designed by Foster+Partners and Renzo Piano Workshop respectively, and have been interested to see how things have gone at SWC.

'What scientists will tell you makes a building work is the science that goes on in there', says Dayan. 'This is going to be a fantastically exciting building to work in. Does the fact that Ian Ritchie's favourite blue is on all the soffits have an impact on our ability to work in the building? That's an interesting question! You'll never know the answer. We've been focusing very much to make sure that the functional aspects of the building work as well as they possibly can.'

As for the experimental research groups, it took longer than envisaged to recruit the first few principal investigators who would bring their teams to the

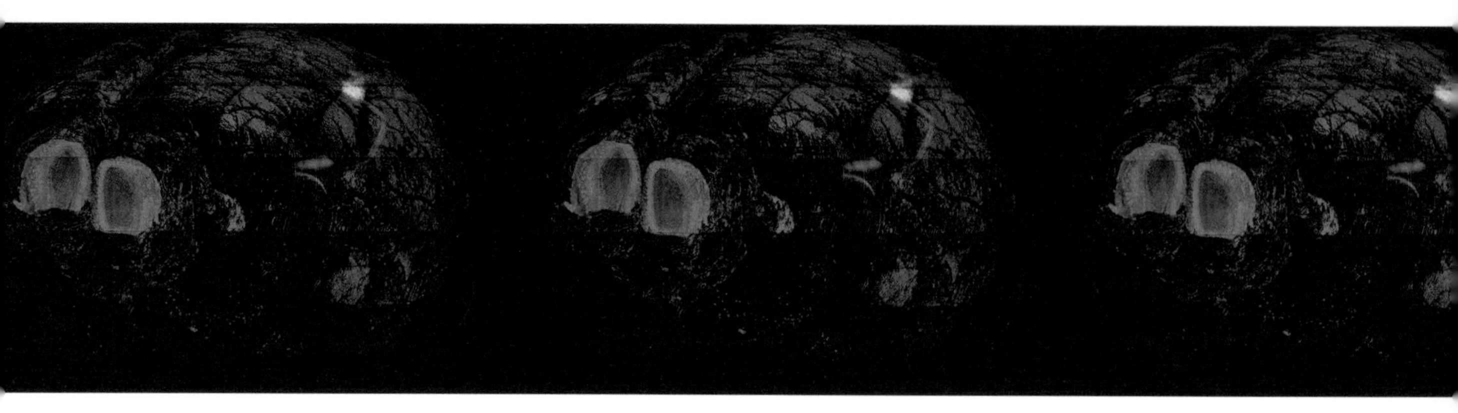

SWC. However beautiful and practical a building, moving a whole research group into an institute that is starting from scratch is potentially a leap in the dark. The slow recruitment was something of a relief for Stuart Johnson as he contemplated a later and later date for the building's handover: 'No one has been left without a lab to go to', he says. The other advantage of having the contractors still on site was that when the first hires were made, they were able to specify alterations to the original design to fit their research needs. 'The building was designed to be adaptable', says Ian Ritchie. 'We had an opportunity to adapt the building before it was even finished.'

The idea was that in each of the four two-tier lab and office spaces, which have come to be known as the 'quadrants', there would be one professorial-level PI plus two early-career group leaders, each with their retinue of postdocs and graduate students. Almost as soon as the base build was complete, John O'Keefe became anxious that with square footage at such a premium in Central London, the four 'mini-atria' between the laboratory floors were perhaps too generous. After a great deal of debate, a section at one end of two atria has been filled in, gaining an extra 16 m^2 of work space per quadrant.

O'Keefe himself would be the senior PI in one of the two lower quadrants. Leading the other was the Australian scientist Troy Margrie, SWC's Associate Director from January 2015, who had previously headed the Division of Neurophysiology at the National Institute for Medical Research (NIMR) in Mill Hill. On 15 April 2015, NIMR became part of the multi-institution Crick Institute: the Crick's new building near St Pancras Station is scheduled to open during 2016. But instead of moving to the Crick's 'cathedral of science', along with another 1500

Brain-wide connectivity mapping

researchers across the biosciences, Margrie chose to come to SWC and become part of a more compact community all working on the same problem.

'We're interested in questions that relate to the influence of experience on the activity of neurons and networks in the brain', says Margrie. He is investigating some of these questions through exploring the way mice use their sense of smell. 'Olfaction is an attractive system in the mouse', he says, 'because the animal is very dependent on that modality. We can train mice to do difficult olfactory-dependent tasks and measure how quickly they can do them, and access the neuronal substrates. And to do those experiments we need a large open space where we can monitor the behaviour of the animals using video. We also need a space where you can close the door and get down to the nuts and bolts of looking at the properties of individual cells.'

Margrie's group also focuses on understanding the complex wiring of the cerebral cortex. 'There are trillions of connections', he says. 'Why does given cell connect to another? How is information being relayed through complex networks?' Modern tracers based on modified viruses enable researchers to follow circuits of neurons across their connections, studying the labelled brain under a two-photon microscope and creating a 3-D image of the connectivity. 'To do that we need a blackout space where no light can penetrate', he says, 'so that we can do measurements of viral particle fluorescence with very sensitive detectors in the microscope.'

The SWC building has all these capabilities – large open space, closed rooms, blackout potential and shared data analysis space all grouped together. Margrie's appointment shifted the emphasis of the work that was likely to be conducted in the building in the direction of an even greater demand for optical techniques. He and O'Keefe quickly agreed that they did not need the 'common support space' that had been designed to occupy a position between the two quadrants on the north side of the building, but did want more spaces that could be blacked out when necessary for fluorescence microscopy and other optical work. With the builders still on site, it was a relatively simple matter to reconfigure the former common support space as a series of closed labs, fitted with floor-to-ceiling blackout blinds inside the glass façade.

What has most impressed Margrie about the SWC's design is the care that has been taken to ensure that laboratory animals are easily accessible to the researchers. 'For me, the most important

HOW DO NEURONS ACT IN CONCERT?
Every neuron is responding, individually, to input from its immediate neighbours, yet collectively they allow us to interact with our environment in a coherent manner. SWC researchers have developed new probes that make it possible to record electrical activity from thousands of neurons simultaneously, which, together with computational analysis, will provide new insights into how populations of neurons act collectively.

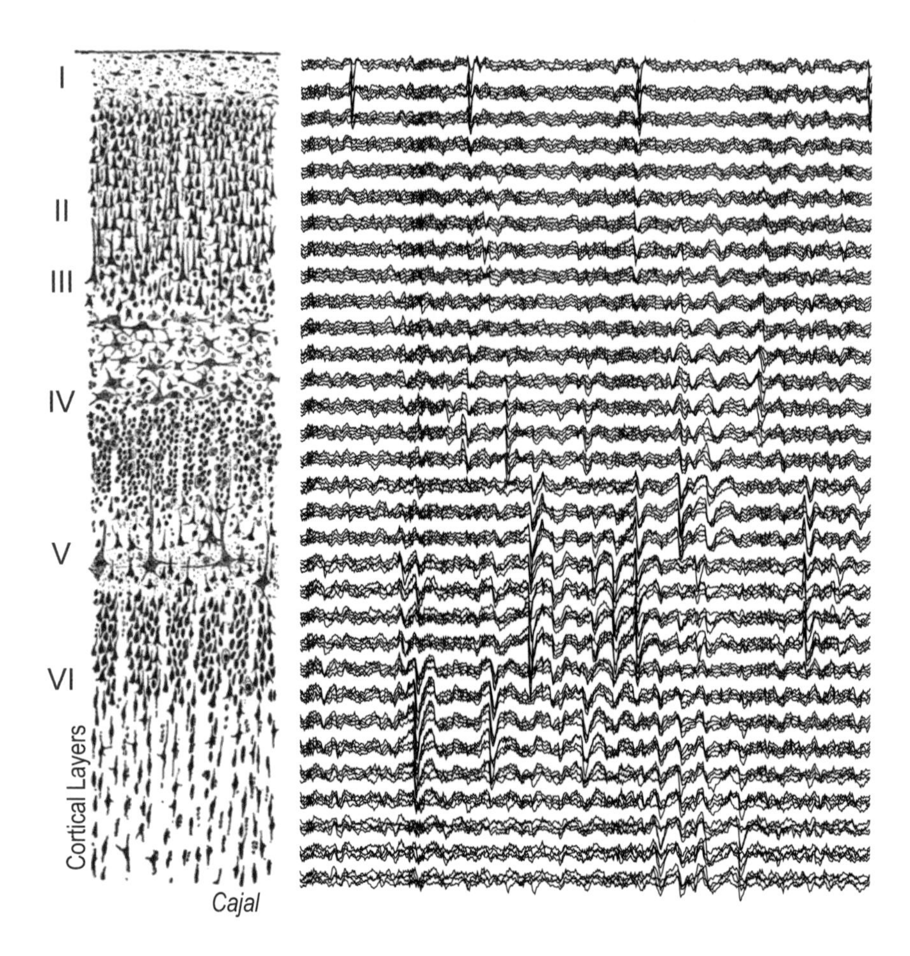

factor in terms of layout is the fact that we have long-term animal holding within our lab environment', he says. 'It means that we can get the answers to our research questions much more quickly than we could previously.'

He is also excited about the fully-equipped fabrication workshop on the ground floor: as Ian Ritchie had found during his original tour of labs, neuroscientists like to make their own kit. 'Traditional mechanical and electrical workshops have been combined with computer hardware development, 3-D printing, additive and subtractive manufacturing', he says. 'To have a large room like this in central London is important, because we can buy large pieces of kit such as 5-axis mills,

Recordings of activity in multiple layers of the cerebral cortex

lathes and laser cutters, and do fine-scale milling, hand lathing, or additive manufacturing. We'll have a core group of four engineers who will interact with scientists to develop projects together. We like to tinker - that's where a lot of the innovation comes from. The commercial suppliers in general follow the scientists, not the other way round.'

For recruitment to more junior positions, SWC cast its net internationally. 'We are seeking the world's best', says Boss. 'That's been facilitated by the building. We have brought potential recruits here, and they are bowled over by the spaces. They in turn became amazing ambassadors for us at the Society for Neuroscience meeting [the annual neuroscience conference in the US, which attracts tens of thousands of participants].' One of these early-career high-flyers is Adam Kampff, an American who did his PhD and postdoctoral research at Harvard, in the basement of the Biological Laboratories that date back to 1932. More recently he had taken up his first PI post at the Champalimaud Centre for the Unknown in Lisbon, Portugal. Arriving at the SWC in September 2015, Kampff immediately appreciated a major difference. 'This building is shockingly good', he says. 'At the Champalimaud, the space where people do experiments is conceptually and optically and physically far away from where they do their other science thinking. What gets me most happy about this place [the SWC] is that the thing that's most annoying about a lot of new developments, they've actually done a good job of solving. Everything feels really connected.'

That connectedness (or 'connexity' to use one of Ian Ritchie's favourite words) is paralleled by the research questions Kampff is looking to address. 'What we mostly do in the lab is explore behavioural space – different challenges that brains are capable of solving – and then we ask what parts of the brain are involved' he says. His team spend much of their time designing 'video games for rats' – real environments in which features can change rapidly by superimposing projected images. 'There are rewards and rules and cues that tell you how well you are doing', he says. 'It can be a racing game, a maze, some sort of different physics where you have to jump from platform to platform. It allows us to change these rules and create a whole lot of new challenges without having to rebuild the setup. And the rats really engage with the game. They will run around and chase spots of light – they have to sneak up on some spots because if you go too quickly they run away. Most of what I would like to understand in rats we can figure out with these contexts.' Other members of his group are designing 'very fancy

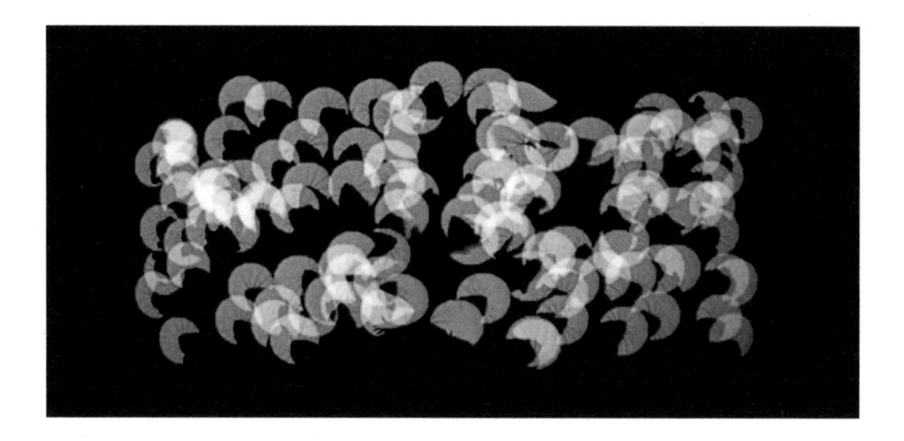

microelectronic-based electrodes', to record from multiple brain structures at the same time. Arriving some months before the labs were equipped and licensed to begin experiments, his junior colleagues occupied their time building equipment in his office.

Looking through the glass wall around the atrium, with its clear view of the labs below, Kampff says, 'As a PI, you are often walking around wondering where your research team are. When they are doing experiments they sometimes disappear for eight or nine hours. Here I can see if one of my postdocs is happy or not when they walk out of their lab room after an experiment, and approach them to find out what they need.' He is relieved that the 'hole in the floor' mostly survived the late push to gain extra square footage on the upper lab levels. 'Without that hole you end up with two separate labs that allow people to hide. Eventually most people said, "Let's keep it there." This atrium was a good idea by the architects in the first place, and I laud them for fighting to keep it, because it's going to define how well the lab works as an entity', he says. 'It creates a collaborative environment without throwing everyone into a great open space and forcing them to deal with it. Here there is a polite, subtle connection between the rooms and the spaces and I am certain that this is going to work very well.'

Andy Murray, a Scot who arrived early in 2016, will be setting up his own research group for the first time, having done his PhD in Aberdeen and worked as a postdoc in Tom Jessell's lab at Columbia before coming to the SWC. He has chosen to study a neural system that often goes unnoticed, but which is essential to our everyday interactions with the physical world: the sense of balance. 'If I

Composite image of rats trained to follow spots of light

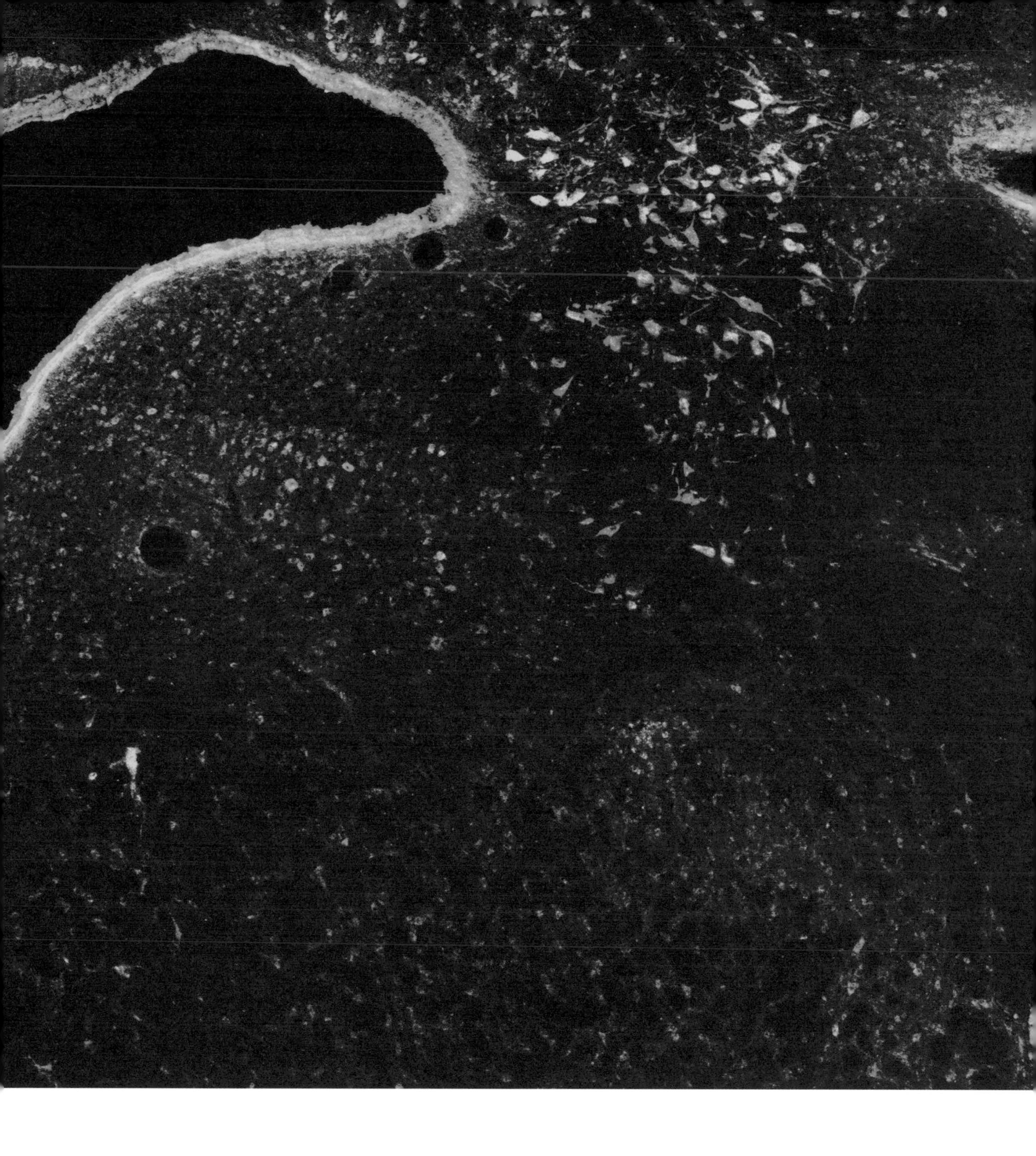

Tracing methods highlight distinct neuronal circuits involved in regulation of balance and stability

were to push you over', he says, 'you would make a very fast movement, putting your arms out, to stop yourself falling. It seems simple, but there's a lot you have to take into account to make a movement quickly and accurately. Otherwise you will fall and hit your head, which is what your nervous system is trying to avoid.'

This sense depends on the vestibular system in the inner ear, which sends output to the lateral vestibular nucleus in the brain stem, which in turn projects to the spinal cord to activate fast reflexes. 'If a mouse is moving along a balance beam, and you move the beam very quickly, the mouse has to compensate for that movement. When we block the vestibulospinal pathway, the mouse can no longer make those quick postural adjustments', says Murray. 'Now I want to look at the information that is going into this pathway, and how that is integrated.'

Murray's approach depends both on measuring behaviour and on using molecular biology to trace pathways, and finds the rooms he's been allocated well set up for both. 'I really like the exposed services', he says. 'One of the best features is that you can have services coming from any point in the room. So if I were to have a behavioural set-up in the middle of the room, I can have a camera and recording equipment coming right down to the middle of the floor. That's a great feature, it's very practical. This is really set up to do science, you get that feel about it.'

As someone whose approach makes use of molecular biology rather than systems neuroscience, Murray was anxious about the shopping list of expensive equipment he would need. But he soon discovered that flow cytometers, centrifuges, gel imaging technology and so on would be bought centrally and shared between all the groups in the building. 'They have really made it so that it is a good molecular biology set-up, as well as systems neuroscience', he says. 'And that's good because neuroscience is really interdisciplinary – you need a range of techniques, from high-level theoretical stuff through systems through to molecular biology and electron microscopy. That's the only way we will ever begin to understand how the brain works. How the equipment and the flow of the building are set up really encourages that broad range of neural circuit analysis.'

The façade, with its translucent cast glass wall transmitting the changing daylight from outside to inside and its opening windows, gets a general thumbs-up. For Adam Kampff, it is just waiting to be used as a blank canvas on which his PhD students

HOW DO NEURONS REPRESENT OUR WORLD?

We know that the brain contains both clocks and maps, otherwise we would not have a sense of time passing nor would we be able to negotiate our environment. These representations are critical to cognitive functions such as memory and planning. Experiments at the SWC will manipulate this perception of time and space in laboratory rodents and human subjects using 'virtual environments', which will help reveal the circuits that underlie these vital systems.

can explore their creativity. 'For the first term the students will build state of the art recording systems and two-photon microscopes, and get a crash course in what it means to do systems neuroscience', he says. 'We can use the distinctive look of the building as an advertising point – this is a different kind of course, and a different kind of research programme.'

Acknowledging possibly difficulties with the planning authorities, he envisages the students deploying the high-powered projectors they use in their research to back-project changing images onto the façade, and for the whole real-time, interactive representation of the building to be reproduced on the SWC's website. 'If you could find a way to control the lighting in the room, you could do all kinds of things using the façade, and if we don't do that then we will have completely missed the point of this building. It's the idea that the place would become a hub,' he says. 'That a crowdsourced, online effort to make sense of brains would be physically instantiated with the cool façade of the building as also the homepage for the website. I think Ian would be disappointed if the building isn't used in this way.'

WHAT DOES THE CORTEX DO?
The cerebral cortex is generally regarded as the 'highest' brain region, and it plays an essential role in intelligent behaviour. Yet it has a surprisingly repetitive modular structure. Cortical circuits are organised into columns, each with six layers, in both mice and humans. How does this columnar unit support intelligent behaviour? Answering such questions would inaugurate a new phase in the science of humans understanding themselves.

For the engineers Arup, Jennifer di Mambro sees the SWC as a prime example of a recent trend in laboratory design. 'It's been quite noticeable over the past 8–10 years', she says. 'Labs have always been a bit of a bugbear – just out-of-town sheds. They were about what goes on inside them rather than enabling the scientists. There have always been aesthetic buildings, like theatres and so on, and process-driven buildings, and now there's a real move to try and combine the two. It will be interesting to see how well that works.'

For Ian Ritchie, the tension between aesthetics and function is to some extent a false dichotomy. 'You can't define the aesthetic intent of a building until you understand what the building needs to do', he says. 'It has to do what the scientists want to be able to do in it, and it has to work for the public realm as well. You might ask, "How do you maintain an aesthetic intent?" when you might be compromised by the scientists' demands, or the planner, or the conservation officer. But I suppose it's the maturity of the designer to take all these balls and juggle them without dropping them.'

This focus on making the building work for its users has let to some truly innovative features, observes Chris Russell of IRAL. 'The double-stacked labs and

The opaque glass façade 'melts' into clear glass
on each corner, giving views into the street below

relationship between lab and write-up – I haven't seen that anywhere else', he says. 'Then having the translucent cast glass façade across the north side is a different approach to a lab. The double layer of glass with translucent insulation exists as a system, but no one's done it toe to toe and modularised it, as far as I'm aware. Then apart from the computational unit and John O'Keefe, we designed the building without knowing the end-users. So we've had a lot of representative users to advise us. And just having so many scientific advisers and lab visits was quite innovative. We were given access to a lot of amazing scientists that we would not have got otherwise, thanks to the influence of Gatsby and Wellcome.'

Stuart Johnson recognises that this level of collaboration owed a great deal to the workshop process that informed the design of the building. 'You can never micromanage average or bad people enough to get them to do something that really works', he says. 'Empowering good people and having the right forum is what's so valuable in terms of getting a fantastic result. I try very hard through setting the ground rules and making sure people understand them, to work in a very consensual way, building a team spirit. Many project managers will ride roughshod over everyone to get an unreasonably cheap, unreasonably quick delivery of the wrong thing. It's not really a response to a brief, because they've never really written one.'

For a modest-sized institute, the SWC began to make an impact internationally even before its laboratories were fully equipped and working. Thanks to Gatsby's flourishing international neuroscience research programme and Ian Ritchie's research-based design process the development of the SWC created a buzz of interest in a world-wide network of neuroscientists. 'Because we consulted so many scientists around the world, these same people are asking how the building's going', says Sarah Caddick. 'I hear them saying, "I talked with the architect, I suggested this". So there's a whole group of people who feel some level of ownership of that building. I think that's not a bad thing.'

Richard Morris, who as neuroscience advisor to the Wellcome Trust helped develop the initial vision for the SWC with Caddick, is well aware that the significance of major discoveries can take years to emerge. In 2016 Morris shared the €1m Brain Prize for his work on the role of synaptic plasticity in learning and memory, work that he began 30 years before. He is confident that the SWC will have an inspiring future. 'There's no reason why the SWC shouldn't crack some of the big problems in systems neuroscience which individual laboratories are unable to do', he says. 'I'd like, in however many years, to be able to say, "Yes, great, we've really made some progress on some big issues."'

John Isaac, a neuroscientist with experience in both academic and industrial research, took over responsibility for the Wellcome Trust's neuroscience programme in 2014. 'The SWC received high priority and large funding', he says, 'because we've seen that these new circuit-busting technologies that really allow us to understand how the brain works on a cellular basis are going to be transformational. At the moment we have blobs of brain lighting up in imaging studies, and we have an understanding of some of the molecules in individual cells, but how they work together to produce brain function is very unclear. This is where SWC is going to be really important.' He thinks the partnership with Gatsby has worked out well. 'The good thing about Gatsby', he says, 'is the care taken in terms of making the building special. It makes people feel special, because they are not just working in an anonymous box on a campus somewhere. It is a bit of a luxury, but my sense is that it pays off in the long run, because people really want to work there.'

David and Susie Sainsbury believe it would be hard to find another scientific research building anywhere that encourages the same level of constant informal interaction between the scientists, and that making all services accessible, so that the 'plug and play' capabilities are genuine, will be immensely valuable. They are confident that Ritchie has delivered a building that permits a high level of practical change to answer the scientists' physical needs, but is also beautiful and architecturally satisfying, and provides a very special working environment.

Watching the distinguished guests tour the SWC on its opening day, Ritchie is confident that it is working in the way he envisaged. 'I think visitors will be very surprised initially', he says. 'They won't have seen a lab like it – not because it set out to be different, but because it set out to be practical and adaptable. I think they'll find it a very delightful space to work in. The philosophy of this office is that we design a building to work for the users – we don't own it, it's not ours. And the conditions that were laid down early – that it must be adaptable, the science comes first, don't waste money on the building – all sound very sensible to us. But at the same time we want something that's aesthetically pleasing to the eye and to the mind. And I think we've got that. The measure, and it's real to us, is that all the work we put in will work for them for a long time. Otherwise there's no point in doing it. Simply no point.'

Above Break-out space on Level 5.
Overleaf Another architectural icon?

TIMELINE

2003

Spring 2003
David Sainsbury meets Sydney Brenner, begins conversations about research on neural circuits

2006

Autumn 2006
David Sainsbury leaves government post as Minister for Science, visits neuroscience labs at Columbia

2007

Summer 2007
Sarah Caddick comes from Columbia to lead Gatsby's neuroscience programme, begins to develop idea for neuroscience institute

Summer 2007
Richard Morris appointed neuroscience adviser to Wellcome Trust, proposes creation of institute for circuits and behaviour

Autumn 2007
Sarah Caddick and Richard Morris meet and propose joint initiative

20 December 2007
Richard Morris submits to Wellcome trustees £50 million proposal for an institute in partnership with Gatsby, which commits £85 million for capital and running costs over the first 5 years

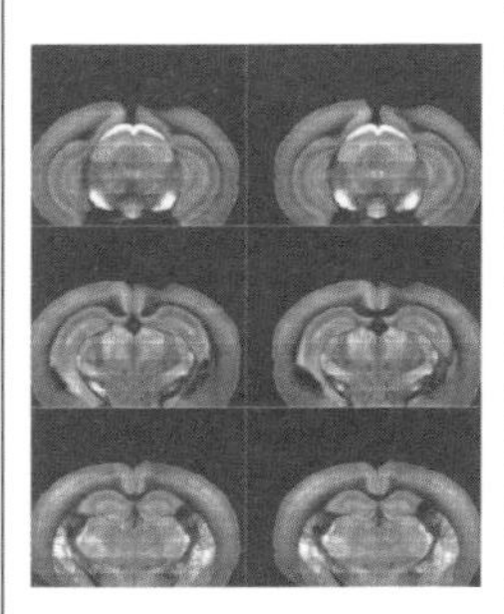

2008

Spring 2008
Wellcome trustees formally accept proposal

November 2008
UCL chosen to host the new institute, site offered on Huntley St

2 December 2008
Meeting between funders, project sponsors and UCL to begin the process of developing a brief for the Sainsbury Wellcome Centre for Neural Circuits and Behaviour (SWC)

2009

July 2009
Briefing document on 'Project Glimmer' completed

September 2009
Agreement signed between Gatsby, Wellcome and UCL to develop the SWC

October 2009
IRAL appointed as architects for 'Project Glimmer'

22 October 2009
Design team interaction day

November-December 2009
Client representatives and architects visit labs in US, UK and elsewhere to investigate how they work and meet neuroscientists

2010

January 2010
IRAL delivers Stage A/B appraisal, outlining problems with Huntley St site

27 January 2010
Neurosciences Interaction Day: 'Educating the architect's team'

March 2010
Windeyer Building on Howland St chosen as site for new centre

4 June 2010
Workshop 2 with scientific advisers – discussions on BSU and vibration control

22 June 2010
Workshop 3 with scientific advisers: options for disposition of BSU, labs, focal space and plant

19 July 2010
Workshop 4 with scientific advisers: review of Stage C options, and vibration workshop

25 August 2010
Workshop 5 with scientific advisers: presentation of Stage C outline proposals, including central position on 3 floors for GCNU, and labs with lower and upper floors

10 September 2010
Design team delivers Stage C report

16 September 2010
Initial meeting with planners from Camden

27 September 2010
Workshop 6 with scientific advisers: sign-off on Stage C and design development aims for Stage D

8 November 2010
Workshop 7 with scientific advisers: Stage D interim review, brief finalised and details of layout finalised

13-14 December 2010
Public exhibition at the Drill Hall

Late 2010
Kier tenders for construction contract

2011

12 January 2011
London Borough of Camden Development Forum

14 January 2011
Workshop 8 with scientific advisers. Recommendation of sign-off on Stage D

28 January 2011
Stage D signed off by Lord Sainsbury for Gatsby Foundation, Sir Mark Walport for Wellcome Trust, and Sir Malcolm Grant for UCL

16–17 February 2011
Interviews held for main contractor

25 February 2011
Planning application submitted to London Borough of Camden

March 2011
Kier is named as preferred bidder and joins team to assist on development of tender information and initial demolition work

13 April 2011
Workshop 9 with scientific advisers: Stage E design development, lecture theatre, catering, adaptability

26 May 2011
Team-building outing to the Royal Shakespeare Theatre in Stratford-upon-Avon

June–July 2011
The Windeyer Building, on the Howland St site, is handed over to the construction team and hoardings erected around it prior to demolition

2012

4 July 2011
Workshop 10 with scientific advisers: Stage E report issued. Final opportunity for members of the Scientific Advisory Council to comment before planning application submitted to London Borough of Camden

13 July 2011
Stage E report signed off by Lord Sainsbury for the Gatsby Foundation; Sir Mark Walport, Director of the Wellcome Trust; and Sir Malcolm Grant, Vice-Chancellor of UCL

21 July 2011
Camden's planning committee accepts planning application. UCL announces that planning permission has been granted the following day

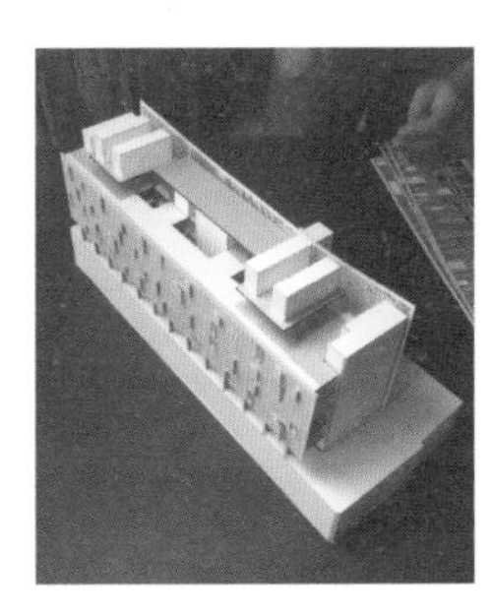

August 2011
Contractors from Kier Group plc begin the demolition of Windeyer Building

5 September 2011
Workshop 11 with scientific advisers. Post-planning and mid-stage F review

23 November 2011
Workshop 12 with scientific advisers. Stage F completion review, update on the detailed design: curtains, floors, furniture etc

29 November 2011–23 January 2012
Scientific Advisory Council members visit laboratory services prototype in Croydon

9 December 2011
Scientific Needs and Equipment Working Group reviews sample equipment layouts

December 2011
The project cost is agreed between Kier, UCL and the funders

January 2012
Demolition of the Windeyer Building completed down to ground level. Basement dig begins, including crushing and recycling of concrete from the demolished building to be reused

8 February 2012
Meeting between Kier and residents of Astor College, a hall of residence of UCL, which borders construction site

8 February 2012
Workshop 13 with scientific advisers: Final cutoff dates for decisions on finishes and furniture agreed. Discussion on catering, sound and light control, public engagement, services, façades and public art

9–10 February 2012
IRAL team visits façade contractor Frener & Reifer to see initial tests of the cast glass panels

February–March 2012
'Room data sheets' produced and circulated by IRAL, so that clients and users have a single source of design information for review

27 February–8 March 2012
Representatives from UCL, contractor and architect visit three shortlisted laboratory furniture manufacturers

21–23 March 2012
Final visits to laboratory services prototype in Croydon

2013

March–April 2012
Two tower cranes erected on site

March–April 2012
Prototype laboratory module in Croydon demolished

18 April 2012
Glass façade tested at Brixen in Northern Italy: tests included the cast glass panels' resistance to impact, and the security of the glazing on the ground floor

May 2012
First of 270 secant piles installed

17 May 2012
UCL issues a press release announcing that contracts 'worth in excess of £70 million' have been signed with Kier to construct the Sainsbury Wellcome Centre, and that construction will begin. Design team novated to Kier

21 May 2012
Workshop 14 for scientific advisers: scientific advisers, client representatives and design and construction teams visit the site for the first time

29 May 2012
Ground-breaking ceremony attended by members of the project team and representatives of the funders and UCL

May 2012
Prototype of the north-facing glass façade constructed for weather testing

June 2012
Excavation of basement begins

10 September 2012
Site Visit 16 for scientific advisors: visit to the construction site to review building progress as well as prototypes of the south façade, laboratory services and write-up desks

26 November 2012
Workshop 17 for scientific advisors: review of finishes to be used in the building

March 2013
Building frame complete to Level 05

June 2013
Façade installation begins

26 June 2013
John O'Keefe announced as the inaugural Director of the SWC

1 July 2013
Topping-out ceremony. Yew tree planted, musicians from the Royal Academy of Music play a piece from Bach's 'Musical Offering'. A representation of the score is to hang in the colonnade in the form of 1000 suspended 'pixels'

4 November 2013
Workshop/site visit 18 with scientific advisers: presentation on art installations for vitrines and projection. Walk through building

December 2013
Façades completed (having been kept partially open for delivery of materials)

2014

11 February 2014
Visit to Luke Hughes & Co to view prototype office furniture

11 March 2014
Workshop/site visit 19 with scientific advisers: building substantially complete. Walk through all levels of building

July 2014
Original target for occupation of building

September 2014
Original target for handover by contractor

September 2014 – November 2015
Unforeseen delays in installation and commissioning of services due to problems with performance of trade contractors

6 October 2014
John O'Keefe announced as winner of the 2014 Nobel Prize for physiology or medicine, jointly with his former colleagues May-Britt and Edvard Moser

2015

January 2015
Troy Margrie from National Institute of Medical Research appointed Assistant Director

June 2015
Director and administrative team move into the SWC

July 2015
GCNU moves into SWC

October 2015
Home Office certifies the building as licensable for the care and use of laboratory animals

24 November 2015
Kier hands over building to clients

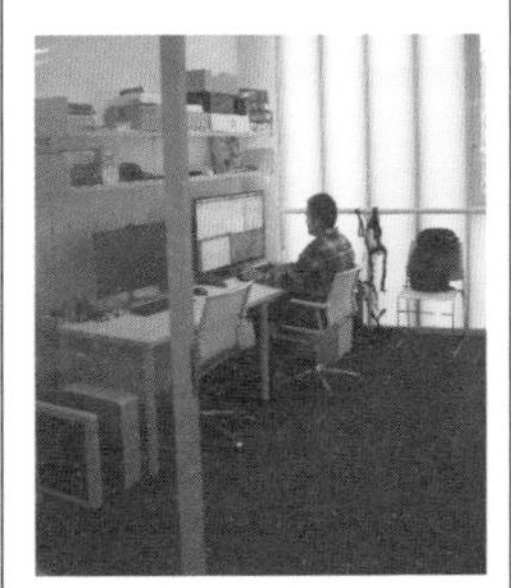

2016

March–April 2016
O'Keefe, Margrie, Kampff and Murray groups move in to their laboratories

May 2016
Isogai group moves in to laboratory

23 May 2016
Inaugural lecture by Eric Kandel and formal opening of building

BUILDING CREDITS

CLIENT
University College London
Professor Sir Malcolm Grant (Provost 2003–2013); Professor Michael Arthur (President & Provost); Professor Sir John Tooke (Vice-Provost (Health) 2009–2015); Professor David Lomas (Vice-Provost (Health)); Professor John O'Keefe (Founding Director, Sainsbury Wellcome Centre)

FUNDERS
The Gatsby Charitable Foundation
David Sainsbury (Settlor), Susie Sainsbury, Peter Hesketh (Director); Sarah Caddick (Neuroscience Adviser)
Wellcome Trust
Sir William Castell (Chairman 2006-2015); Lady Manningham-Buller (Chair); Sir Mark Walport (Director 2003–2013); Professor Jeremy Farrar (Director); Dr Kevin Moses (Director of Science); John Williams (Head of Neuroscience and Mental Health 2008–2015); Dr John Isaac (Head of Neuroscience and Mental Health)

PROJECT SPONSOR AND FUNDERS' REPRESENTATIVE
Stuart A Johnson Consulting Limited
Stuart Johnson

PROJECT AND CONTRACT ADMINISTRATOR
Peter Brett Associates LLP (formerly Hannah-Reed and Associates)
Eamonn Rogan (Associate Project Manager); Ron Henry; Malcolm Fillingham; Henry Martin; Margaret Winchcomb; Emel Kus; Mark Hall

UCL ESTATES AND FACILITIES
University College London
Dave Smith (Deputy Director, Directorate); Richard Bayfield (Development and Projects Executive, Capital Programmes and Procurement)

USER REPRESENTATIVE
Wolfson Institute for Biomedical Research at UCL
Arifa Naeem

ARCHITECT
Ian Ritchie Architects Ltd
Ian Ritchie (Director); Gordon Talbot (Projects Director); Chris Russell (Project Architect); Anthony Summers (Projects Director); Shee Ming Loke; Teresa McQueen; Dana Bilek; Brian Heron; Karl Singporewala; Akhil Bakhda; Annamaria Csanaki; Ben Dudek; Sara L'Espérance; Rossa Prendergast; Gordon Swapp; Ben Ward; Jerome Berteloot; Kieran Wardle; Andrew Hum

STRUCTURAL, CIVIL AND BUILDING SERVICES ENGINEERING
Arup
Andrew Harrison (Director, Project Director); Jennifer DiMambro (Director, Project Lead); Christian Allison (Director, Structural Lead); Michael Lorimer (Associate Director, Building Services Lead); Arnon Dienn: Bert Fraza; Paul Jeffries; Hugh Pidduck; David Bloomfield; Stuart Jordan; Michael Lorimer; Morwenna Wilson; Gary Reid; Pamela Nwaneri; Mike Ebsworth; Greg Howard; Jenny Bousfield; Katie Bearder; Yeside Sogunro; David Cologne; Stuart Hitchcock; Julie Howes

ACOUSTICS
Arup
Philip Wright

BREEAM ASSESSOR
Ian Ritchie Architects Ltd
Chris Russell

COST CONSULTANT
EC Harris LPP
Mark Cleverly (Partner); Robert Littlewood (Associate); Colin Duke; Paul Campbell; Bryony Day; Alicia Nevin

LANDSCAPE ARCHITECT
Ian Ritchie Architects Ltd
Hoo House Nursery
Ian Ritchie (Director); Robin Ritchie (Director)

LIGHTING CONSULTANT
EQ2 Light
Mark Hensman (Director)

LABORATORY PLANNING
David Kelly Associates
David Kelly

MAIN CONTRACTOR
Kier Major Projects
Colin Lamb (Managing Director Major Projects); Nick Mann (Senior Project Manager)

FIRE
Arup
Hay-Sun Blunt

SECURITY
Arup
Jeff Green

LOGISTICS
Arup
Alan Beadle

VIBRATION
Arup
James Hargreaves

AV
Arup
John Sykes

CDM COORDINATOR
Peter Brett Associates LLP [formerly Hannah-Reed and Associates]
Graham Fowler

ACCESS/DDA
Centre for Accessible Environments
Keith Garner

HERITAGE CONSULTANT
Peter Stewart Consultancy
Peter Stewart

PLANNING CONSULTANT
Dp9
David Graham

PARTY WALL CONSULTANT
Goodman Mann Broomhall
Henry Watson

RIGHTS OF LIGHT CONSULTANT
Gordon Ingram Associates
Gordon Ingram

TRANSPORT CONSULTANT
Peter Brett Associates LLP
[formerly Hannah-Reed and Associates]
James Martin; Paul Cosford; Kushal Patel

ECOLOGY CONSULTANT
Ecological Planning & Research Ltd
Gary Cole

EMC CONSULTANT
Technology International (Europe) Ltd
(Cobham Technical Services - Stages A-C)
Alex McKay

ARCHAEOLOGY CONSULTANT
Museum of London Archaeology
Nick Bateman

NEC3 PROJECT SUPERVISOR TEAM
Peter Brett Associates LLP
(formerly Hannah-Reed and Associates)
Andrew Durham; John Collier;
Max Roach

IT CONSULTANT
Cordless
Nigel Miller

CATERING CONSULTANT
Harbour and Jones
Patrick Harbour; Nathan Jones

GRAPHICS AND BRANDING
Scott Thornley + Company
Scott Thornley

BUILDING CONTROL
London Borough of Camden
Peter Connell

FURNITURE CONSULTANT AND CONTRACTOR
Luke Hughes & Co
Luke Hughes

ARTISTS' VITRINES
Marty Banks Consulting
Marty Banks, Hany Farid, Maria Mortati

COLONNADE PIXELS
Ian Ritchie Architects Ltd

PROJECTION WEST WALL
Pani (Projectors)

KIER SUBCONTRACTORS
Piling
Bachy Soletanche Ltd
Frame
A J Morrisroe & Sons Ltd
Structural Steelwork
Elland Steel
Cladding
Frener & Reifer
Wall Cladding [Astor College]
ECL Contracts
Roofing
Fenland Flat Roofing
MEP
Balfour Beatty Engineering Services [BBES]
Brick & Blockwork
Grangewood
Isolated Slabs
Mason UK Ltd
Joinery
Brown & Carroll
Screeds & Drylining
T Lott
Decorations
FH Harvey
Flooring
AC Flooring
Hard Landscaping
Maylims
Soft Landscape
Goddards
Irrigation
Waterscapes Ltd
Architectural Metalwork
Glazzard Ltd
Internal Glazed Partitions
Planet Partitioning
Internal Hygienic Doors
Dortek
Internal Steel Doors
Allegion; Ingersoll Rand Martin Roberts
Ironmongery
Allgood
Lifts
Orona
Turntable
Movetech
Lab Furniture
WE Marson & Co Ltd
Lab Equipment
Getinge UK Ltd; Tecniplast UK; Bell Isolation
Security Portals
Meesons
Kitchen Fit-Out
IFSE
Roller Shutters
Hormann
Movable Partitions
Style South
Blinds
Solar Contracts
Signage
DMA Signs
Manifestation
Window Film Company
UKPN Doors
Tegrel

IMAGE CREDITS

Grant Smith
pp. 2–3,7,8, 88, 87, 89, 91, 93, 94, 98, 99, 103, 104, 105, 106, 107, 113, 119, 126–7, 129, 130, 132, 134, 137, 138, 144–5, 148, 156–7, 170, 173, 174–5, 176, 181 (far left), 181 (far right)

Lord Sainsbury
p. 6

Wellcome Trust
p. 9

Axel laboratory, Columbia University
p. 16

Gretton laboratory, GCNU
p. 22

IRAL
pp. 28, 40, 42, 45, 46, 49, 52, 54, 56, 57, 58, 59, 60 (left and right), 63, 64, 65, 66, 72, 73, 74, 75, 76, 78–79, 81, 82, 87, 101, 114, 119, 121, 143, 146, 149, 151, 177 (right), 178 (far left), 178 (second left), 178 (right), 179 (left), 180 (far left), 180 (second left), 181 (second left), 181 (second right), 189

Häusser laboratory, UCL
pp. 31 and 43

Daniel Berger, Seung and Lichtman laboratories, Harvard
p. 35

Waddell Laboratory, Centre for Neural Circuits and Behaviour, University of Oxford
p. 36

Mary Hinkley/UCL Digital Media Services.
p. 155

Margrie laboratory, SWC
pp. 162, 177 (left)

Kampff laboratory, SWC
pp. 164 and 166

Murray laboratory, SWC
p. 167

UCL
p. 177 (centre)

Kier Construction
pp. 179 (right) and 180 (right)

Carl Bigmore
pp. 136 and 138

Adam Smith
p. 139

VISUAL ILLUSIONS FOR CHAPTER OPENINGS
We see illusions because our brains make bets on what we are seeing based on previous experience.

Chapter 1 (p 10)
Are the horizontal lines parallel? In the 'café wall illusion' we see alternate vertical lines slope down to right and left, when in fact they are parallel. For the illusion to work the 'mortar' in the wall must be intermediate in lightness between the dark and light squares.

Chapter 2 (p 38)
How many shades of blue? In White's illusion our perception of lightness/darkness is distorted by the background. We see three shades of blue, whereas in fact there are only two.

Ian Ritchie, architect

ACKNOWLEDGEMENTS

Chapter 3 (p 68)
How many black dots? In this scintillating grid illusion there appear to be black and white dots at the intersections, but if you focus on any intersection the black dot disappears. There are, in fact, only white dots.

Chapter 4 (p 84)
Are the diagonal lines parallel? The Zöllner illusion demonstrates how intersecting short lines fool us into thinking the longer lines are nearer to us at one end than the other, giving the illusion that they diverge.

Chapter 5 (p 120)
How many shades of blue? Another effective example of White's illusion. There is also an illusion of depth: the 'darker' triangles appear to be behind the dark blue stripes, while the 'lighter' triangles appear to cross in front of them.

Chapter 6 (p 152)
Which way do the arrows go? An example of a bistable image: you either see the light arrows pointing upwards against a dark background, or the dark arrows pointing down against a light background, but you can't see both at the same time.

Illustrations by Kathrin Jacobsen

Endpapers
Dreaming of a Project,
Ian Ritchie RA 2009.

The author would like to express her gratitude to all those who helped in the compilation of this book, including many of the scientific advisers, design team, client team and construction team who are named in the text. She would especially like to thank Susie Sainsbury and Ian Ritchie, whose conception the book was and who have been unfailingly supportive, and Kathrin Jacobsen, the designer, who was a joy to work with. Any errors are entirely the author's responsibility.

GLOSSARY/INDEX

A

B

D

F

G

H

I

J

K

L

M

N

P

Q

R

S

T

U

Z